National Guard Youth ChalleNGe

Program Progress in 2018–2019

LOUAY CONSTANT, JENNIE W. WENGER, LINDA COTTRELL,
STEPHANI L. WRABEL, WING YI CHAN

Prepared for the Office of the Secretary of Defense
Approved for public release; distribution unlimited

For more information on this publication, visit www.rand.org/t/RR4294

Library of Congress Cataloging-in-Publication Data is available for this publication.
ISBN: 978-1-9774-0449-7

Published by the RAND Corporation, Santa Monica, Calif.

Cover photo: Alex Rafael Román/Prensa LA FORTALEZA.

Cover design: Katherine Wu

Preface

The National Guard Youth ChalleNGe program is a residential, quasi-military program for youth ages 16 to 18 who are experiencing difficulty in traditional high school. This report covers the program years 2018–2019 and is the fourth in a series of annual reports that RAND Corporation researchers have issued over the course of a research project spanning September 2016 to June 2020.[1] The first, second, and third National Guard Youth ChalleNGe Annual Reports cover program years 2015–2016, 2016–2017, and 2017–2018, respectively, and can be found on the RAND Corporation website (Wenger et al., 2017; Wenger, Constant, and Cottrell, 2018; Constant et al., 2019).

Each annual report documents the progress of participants who entered ChalleNGe during specific program years and then completed the program. A focus of RAND's ongoing analysis of the ChalleNGe program is collecting data in a consistent manner. Based on these data, each report also includes trend analyses. In this report, we provide information in support of the National Guard Youth ChalleNGe Program's required annual report to Congress. In addition to information on participants who entered the ChalleNGe program and completed it in 2018, we include follow-up information on those who entered the program and completed it in 2017. Finally, we describe and provide syntheses of other ongoing research efforts to support the ChalleNGe program. Full details of these efforts will be documented in greater detail in the future.

Methods used in this study include site visits, collection and analyses of quantitative and qualitative data, literature reviews, and development of tools to assist in improving all program metrics—for example, a program logic model. Caveats to be considered include some documented inconsistencies in reported data across sites, our focus on those who completed the program and not on all participants, and the short-run nature of many of the metrics reported here.

This report will be of interest to ChalleNGe program staff and to personnel providing oversight for the program. This report may also be of interest to policymakers and researchers concerned with designing effective youth programs or determining appropriate metrics by which to track progress in youth programs. The research reported here was completed in November 2019 and underwent security review with the sponsor and the Defense Office of Prepublication and Security Review before public release.

This research was sponsored by the Office of the Assistant Secretary of Defense for Manpower and Reserve Affairs and conducted within the Forces and Resources Policy Center of the

[1] This report draws heavily on information initially provided in our previous reports (Constant et al., 2019; Wenger, Constant, and Cottrell, 2018; Wenger et al., 2017).

RAND National Security Research Division (NSRD), which operates the National Defense Research Institute (NDRI), a federally funded research and development center sponsored by the Office of the Secretary of Defense, the Joint Staff, the Unified Combatant Commands, the Navy, the Marine Corps, the defense agencies, and the defense intelligence enterprise.

For more information on the RAND Forces and Resources Policy Center, see www.rand.org/nsrd/frp or contact the director (contact information is provided on the webpage).

Contents

Figures

Tables

Summary

The National Guard Youth ChalleNGe program is a residential, quasi-military program for youth ages 16 to 18 who are experiencing academic difficulties and exhibiting problem behaviors inside and/or outside school, have either dropped out or are in jeopardy of dropping out of their high school, and, in some cases, have had run-ins with the law. The ChalleNGe program is 17.5 months in length, broken into a 5.5-month Residential Phase (comprising a two-week acclimation period, called Pre-ChalleNGe, and the five-month ChalleNGe) followed by a 12-month Post-Residential Phase. During the Post-Residential Phase, graduates may continue their education, find employment, enlist in the military, or undertake some combination of these. Each graduate has a mentor whose role is to provide advice, assist with the transition after ChalleNGe, and provide monthly reports back to the program about the graduate's placement (i.e., education, employment, military). Graduates and mentors are expected to meet regularly.

Participating states operate the program, which began in the mid-1990s, with supporting federal funds and oversight from the state National Guard organizations. There are currently 39 sites in 28 states, the District of Columbia, and Puerto Rico. More than 230,000 young people have taken part in the ChalleNGe program, and close to 175,000 have completed the program.

ChalleNGe's stated mission is to "intervene in and reclaim the lives of 16–18-year-old high school dropouts, producing program graduates with the values, life skills, education, and self-discipline necessary to succeed as productive citizens" (National Guard Youth ChalleNGe, 2015, p. 2). The program delivers a yearly, congressionally mandated report documenting progress; the data and information in this report support this requirement.[2] Previous research on the ChalleNGe program has found that it is cost-effective and has positive effects on the educational and labor market outcomes of participants (referred to as *cadets*): ChalleNGe participants achieve more education and have higher earnings than similar young people who do not attend the program (see Bloom, Gardenhire-Crooks, and Mandsager, 2009; Millenky, Bloom, and Dillon, 2010; Millenky et al., 2011; and Perez-Arce et al., 2012).

The ChalleNGe program emphasizes the development of eight core components: leadership and followership, responsible citizenship, service to community, life-coping skills, physical fitness, health and hygiene, job skills, and academic excellence. There is variation across the 39 ChalleNGe sites in the variety of program activities implemented to support the program's core

---

[2] This RAND research draws heavily on the previous three reports in the series (Constant et al., 2019; Wenger, Constant, and Cottrell, 2018; Wenger et al., 2017). The previous reports include additional background information and detail on the ChalleNGe program and model. The previous reports also include a more detailed description of the development of the logic model and additional information on standardized test scores used and tracked within the ChalleNGe program.

components. The factors that determine this variation are a combination of state and local context, program history, and site-level preferences. The ways in which site-level variations might result in different levels of long-term effectiveness are not known. For example, one key area of variation across program sites is the type of academic credential offered; many cadets work toward a General Education Development (GED) or High School Equivalency Test (HiSET) certificate, but some sites allow cadets to work toward a high school diploma or collect high school credits that allow them to transfer back to and graduate from their home high schools after returning home. Although sites tend to emphasize one approach over another, many sites offer multiple options for cadets, and a cadet's choice typically will depend on his or her age, credit status at the time of enrollment, and personal and/or family preferences. Additionally, some sites offer specific occupational training or certificates, and some offer the opportunity to earn college credits. In another example, although all sites provide volunteer opportunities for ChalleNGe cadets to fulfill the core component of providing service to the community, the types of opportunities differ based on the nature and range of partnerships that individual sites have developed. In addition to the two core components of academic excellence and service to community, sites vary on several other aspects of implementing the remaining six core components.

Objectives of the Project

The RAND Corporation's ongoing analysis of the ChalleNGe program has two primary objectives. The first is to gather and analyze existing data from each ChalleNGe site on an annual basis to support the program's yearly report to Congress. This RAND report is the written product of the research team's fourth round of data collection. As in our first, second, and third annual reports, we requested information from each ChalleNGe site. The core metrics collected have remained more or less consistent across the years. The bulk of the analysis in each annual report to Congress is based on these data. We collected measures that include tallies of the total number of young people who participated in and successfully completed the program, the number and type of credentials awarded, standardized test scores, participation in community service, and registration for voting and Selective Service. Each annual report also includes detailed site-specific information on participants, staffing, and funding. Finally, each annual report includes information on post-ChalleNGe placement (for example, enrollment in school, participation in the labor market, or enlistment in the military). We present measures at the site level; in some cases, we include aggregated information as well.

The second objective of this project is to identify longer-term metrics for the overall effectiveness of the program, including ones that will help determine how site-level differences influence program effectiveness. The relevant information comes from multiple sources, including measures collected for the annual report, analysis of qualitative data collected from site visits, and analysis of extant data, such as national samples matching the profile of the population of interest. In this report, we continue to document our progress toward this objective. For example, we have undertaken several analytic tasks and two pilot projects in close collaboration with ChalleNGe sites. (For organizational purposes within this project, we define *analytic tasks* as research efforts that use information from multiple ChalleNGe sites; *pilot projects* are research efforts that involve working closely with a single site to implement a program based on best-practices. Both types of efforts could have implications for all ChalleNGe sites). In this report, we share a synthesis of findings from one analytic task and from one pilot project. The analytic task is examining Career and Technical Education (CTE) practices in program sites, and the pilot is designing a mentor training initiative drawing on best-practices from the mentoring literature. Note that analytic tasks can lead to a pilot. For example, we previously provided a synthesis of the mentoring literature and findings on mentoring from site visits (Constant et al., 2019), which was then used to design the mentoring pilot described in this report.

We also describe two other analytic efforts being undertaken—one on the regional return to occupational credentialing and the other on efforts of ChalleNGe sites to meet the mental health needs of their cadets. These two analytic efforts are currently in their early stages, and thus we do not share findings in this report. Finally, we describe another pilot study that consisted of partnering with a ChalleNGe site to design and implement a ChalleNGe alumni survey to collect placement information from graduates of that site.

Cross-Site Measures for the 2018 ChalleNGe Classes

The quantitative information included in this report was collected from ChalleNGe sites in July and August 2019, using the same approach described in previous reports (see Wenger et al., 2017; Wenger, Constant, and Cottrell, 2018; Constant et al., 2019) but with refinements based on project needs and feedback from the program sites.

It is important to note that schedules vary somewhat across ChalleNGe sites; we requested that each site send information on the cadets who entered the program in 2018. Typically, there are two cohorts of cadets in a year per site, with notable exceptions.[3] The two classes described in this report are generally referred to by program staff as Class 50 and Class 51 (ChalleNGe classes are numbered consecutively from the first class in the 1990s).[4]

Our current findings reveal that Classes 50 and 51 produced 9,352 ChalleNGe graduates. Around 43 percent of graduates received an official credential—either a high school diploma or a credential based on passing a standardized test. When we include high school credits as well, about 60 percent of cadets received at least one academic credential. These figures are similar to those reported in previous years. As a group, these graduates performed more than $14 million worth of service to their communities. The overall graduation rate for these two classes was 73 percent, based on the number of cadets who enter the two-week acclimation period termed Pre-ChalleNGe.

To date, we have collected information on eight classes of ChalleNGe cadets; our previous reports include information on Classes 44 to 49 (Wenger et al., 2017; Wenger, Constant, and Cottrell, 2018; Constant et al., 2019). As a result, we can examine trends over time. Our trend analysis indicates that ChalleNGe programs took in fewer participants in 2018 than in 2017 (down from 13,457 to 12,844). Some of the decrease is due to programs consolidating (e.g., Texas) without the addition of new programs, as was the case from 2016 to 2017. But another factor is a decline in the size of the programs; this decline is concentrated among the largest programs. The underlying reasons for the decline are unclear, but site visits suggest that

[3] At Wyoming Cowboy ChalleNGe, three cohorts began during 2018; cadets were enrolled in two overlapping cohorts in the first half of the year, but the program shifted to a more traditional two-cohort-per-year schedule in July 2018.

[4] There are exceptions, particularly with sites where classes cross over calendar years. The class numbering maintained by the program site may not correspond with the class numbering referred to by the National Guard Bureau.

some ChalleNGe programs face more competition than before and face difficulties in meeting their recruiting targets. In terms of efficiency, the effects of this decline are likely to be small; our high-level estimate of program costs by size indicates that the smallest programs are more costly per participant than larger programs, but per-cadet costs are similar across most programs (Wenger et al., 2017).

The overall graduation rate remained roughly consistent with prior classes. Due to recent changes in the standardized test (the Tests of Adult Basic Education, or TABE)[5] used by ChalleNGe sites, we do not compare trends in test scores. However, among sites that have not yet adopted the updated test, scores reported this year are similar to those reported in past years. Thus, academic quality appears to be roughly constant over the study period.

Tests of Adult Basic Education Scores

Recently, the TABE has been updated from TABE 9/10 to a new version, TABE 11/12, to reflect changes in educational standards based on the 2010 release of the K–12 education Common Core State Standards (CCSS), which led to the identification of new College and Career Readiness (CCR) standards for Adult Basic Education (ABE). The TABE 11/12 was created to align with these standards. While most ChalleNGe sites continued to use the TABE 9/10 in 2018, some sites adopted the TABE 11/12. In the near future, all sites will begin using the TABE 11/12. Scores from the TABE 9/10 and TABE 11/12 are not comparable, and thus TABE scores from sites that administered the 9/10 cannot be combined with TABE scores from sites that administered the TABE 11/12. For the purposes of this report, information on the TABE is presented separately by test version. Moreover, we cannot include scores from sites that administered the TABE 11/12 in trend analysis across classes because scores from previous years are based on the TABE 9/10.

Among sites that continued to use the TABE 9/10, analysis revealed substantial progress over the course of the Residential Phase, with one-third of cadets scoring at the ninth-grade level or higher at the beginning of the Residential Phase compared with more than half of cadets scoring at or above the ninth-grade level by the end of the Residential Phase. In terms of the sites that took the TABE 11/12, by comparison to sites that took the TABE 9/10, scores are lower. Less than 15 percent of cadets initially scored at or above the ninth-grade level; by the end of the Residential Phase, less than one-quarter scored at the ninth-grade level. TABE 11/12 scores indicate growth over the course of the Residential Phase, but comparisons suggest that cadets will score lower on the TABE 11/12 compared with the TABE 9/10. Since the TABE 11/12 has not yet been linked to outcomes of interest, such as the passing rate on the GED subject sections in the same way that the TABE 9/10 has, these results should be interpreted with caution. Scoring lower on the TABE 11/12 than on the TABE 9/10 does not necessarily imply a decline in academic achievement.

[5] The TABE is a standardized assessment of adult basic education that tests math, reading and language arts, social studies, and science knowledge and skills for adult learners. ChalleNGe sites are required to test incoming cadets in math and reading twice during the Residential Phase, typically at the beginning and then again closer to graduation, to assess progress. For more information on TABE, see tabetest.com.

Staffing

The study team also collected information on the number of staff by position, as well as the number of staff by position newly hired in the past 12 months, as an indicator of staff turnover. At a typical site, roughly 15 percent of administrators and instructors have been employed for less than 12 months. Among cadre, who are the staff that oversee the quasi-military aspect of the program and spend most of their time overseeing the cadets, the turnover rate is twice as high, with a typical site reporting that about 30 percent of cadre have been employed for less than 12 months.

While there could be numerous factors contributing toward high staff turnover, one possible factor is the starting salary. Comparing program sites with less than 50 percent of cadre hired in the past 12 months as an indicator of lower turnover with program sites with more than 50 percent of cadre hired in the past 12 months, we found that programs with lower turnover pay their cadre starting salaries that are, on average, $6,000 higher. Similarly, for instructional staff, sites with less than one-third of their instructors hired in the past 12 months pay their instructors starting salaries that are, on average, $7,000 more than sites that have hired one-third or more of their instructional staff in the last 12 months. While salary is not the only factor that may be influencing turnover, it is one area worthy of further exploration.

Placement

The study team also examined Post-Residential placement data collected and reported by the program sites. By months 6 and 12, and similar to the 2018 data, nearly 80 percent of graduates are listed as having a placement, with education and employment as the most common placements. Cadets are most likely to be enrolled in school in the first month after graduation; in later months, cadets are more likely to be employed. The proportion of cadets who report military service increases over the months, as does the proportion who report some combination of education, employment, and military service. Notably, programs struggle to obtain placement information on all cadets. Among Classes 50 and 51, sites reported information on 75 percent of cadets at the six-month mark and 67 percent at the 12-month mark.

Time Trends, 2015–2018

With four years of data reported by the program sites, it is possible to show time trends on some key indicators, including total number of participants, graduates, and graduation rates. Trends reveal that the number of cadets participating in ChalleNGe increased slightly from 2015–2017, but then there was a decline from 2017 to 2018. The graduation rate has remained constant, but the absolute number of graduates declined from 2017 to 2018. Because some programs adopted the TABE 11/12 during 2018, we do not report trends in TABE scores. But we do note that among programs continuing to use the TABE 9/10 throughout 2018, scores were quite similar to those recorded in earlier years.

Analytic Efforts

In addition to preparing this year's annual report, the fourth in the series, the RAND study team also undertook several analytic efforts that address components of the National Guard Youth ChalleNGe program and relate to the program's logic model. The logic model describes the design of the National Guard Youth ChalleNGe program, the activities that different program sites implement, and the expected outputs and outcomes. Our analytic efforts are intended to address gaps in data collection, particularly around long-term outcomes, and better understand program design and implementation issues (for instance, how to improve the mentoring component).

In 2019, we reported on two analytic efforts, one on mentoring and the other on benchmarking youth outcomes (Constant et al., 2019). In this report, we review a few of the analytic efforts that we developed over the past year in support of the ChalleNGe program. The research draws on best-practices from the broader literature and includes research on a variety of topics relevant to the program's eight core components, particularly academic excellence and job skills, life-coping skills, and health and hygiene. To address these core components, we report on analytic efforts related to CTE, mental health supports, and occupational credentialing. In some cases, we employ empirical methods to analyze data relevant to ChalleNGe. For example, using publicly available data from nationally representative surveys, the RAND research team is examining the regional payoff of specific credentials in the occupational credentialing study; this information will be useful as the program considers which credentials should be offered and emphasized. We also draw on practices in the program sites and the challenges they encounter—for example, we examine CTE provision and the measures that ChalleNGe sites have taken to meet the mental health needs of cadets enrolled in their programs. Below, we describe each of the three analytic efforts in more detail.

Examining the Implementation of Career and Technical Education

In this analytic study, the RAND team focused on examining CTE practices across ChalleNGe sites. The RAND team sought to identify promising practices in CTE and understand both the opportunities and constraints to providing CTE at ChalleNGe sites. The RAND team examined the current literature on CTE to identify promising practices; described CTE practices across ChalleNGe sites by drawing on the data reported in the annual data call and benchmarking those practices with national data; and conducted interviews with select ChalleNGe sites to gain a deeper understanding of CTE implementation.

A review of the literature suggested five promising practices in the implementation of CTE: (1) offering *structured pathways*, or a sequencing of courses and training that place students on a defined path toward earning a credential; (2) providing *career preparation supports* to help students align their course-taking with future career plans; (3) participating in *work-based learning (WBL)* opportunities to apply skills in formal work-based settings; (4) promoting an *integrated academic-occupational curriculum* that explicitly links concepts and skills across academic and occupational learning materials to promote both college and career readiness; and (5) investing in *industry engagement* to maintain relevant occupational offerings, updated curricula, and WBL opportunities for students. Recognizing that ChalleNGe sites face constraints in applying these promising practices to their specific circumstances, the study team sought to highlight areas where sites are finding some success in doing so.

Data collected on CTE participation and offerings across sites suggest that there is wide variation in CTE participation. A handful of sites could be considered CTE intensive sites where CTE participation among cadets closely matches CTE participation in traditional public high schools in the United States. Not surprisingly, ChalleNGe sites tend to offer courses that are much more occupationally driven than traditional public high schools, which are more explicitly preparing high school students for four-year colleges.

Focused interviews with a handful of ChalleNGe sites to understand the opportunities and challenges to offering CTE suggest that sites struggle to integrate CTE into cadets' schedules due to the intense focus on completing their academic requirements. This is particularly the case with sites that have to transport their cadets to participate in CTE, which is typically done by enrolling cadets in dual enrollment programs in local community colleges. While some sites make CTE-type courses available to all cadets, sites have also developed eligibility criteria for certain CTE classes that are only offered to cadets who are progressing on schedule in terms of meeting their academic requirements. The expansion of Jobs ChalleNGe, though not to all sites, has nonetheless prompted sites to consider ways in which they might better align current and future CTE offerings with similar types of postresidential training opportunities in subbaccalaureate fields.

Examining Mental Health Supports

Because of concern over the growth of mental health issues among adolescents, and particularly among those who are considered at-risk, there is a need for an evidence base on what works to guide interventions. In the case of ChalleNGe, interviews conducted during visits to program sites suggest that some cadets have previously diagnosed mental health conditions that require medication and/or counseling. These conditions may or may not be disclosed during the application process; also, cadets can have undiagnosed mental health conditions that emerge during their time at ChalleNGe. The current research seeks to develop a set of recommendations for best-practices to address the mental health needs of cadets. The team will conduct a review of the literature on best-practices for identifying and treating adolescents with mental health concerns; interview a set of counselors at ChalleNGe academies to better understand their policies and procedures, staffing models, and innovations being considered and/or applied; and develop questions for the annual data call to assess models currently in place across all ChalleNGe programs.

Examining Occupational Credentialing

The purpose of this analysis is to identify the skills prevalent among occupations that are high-paying, high-growing, or both, so that sites can incorporate skill-training, if not job-training, into their curriculum. The occupation report starts by identifying "good jobs" for workers without a college degree, and then determining the skills, abilities, and job features among these jobs. This information will assist ChalleNGe sites as they establish job-training and life-skills curriculum by providing specific examples of skills and abilities associated with labor market rewards. The final report will present skill summaries of good jobs by levels of education (such as high school diploma holder versus occupational certificate holder), and by industry (such as construction versus sales).

Pilot Projects

The RAND team has also helped two program sites implement two distinct pilot projects. In one case, the pilot project drew on best-practices from a review of the mentoring literature published in a previous report (Constant et al., 2019) to design a mentor training pilot suitable to a ChalleNGe site. In this report, we describe the design of the pilot and share early findings from implementing it in one of the ChalleNGe sites. We also briefly describe the design of an alumni survey pilot in an effort to collect information past the Post-Residential Phase on ChalleNGe graduates from one program site.

Improving Mentor Training

The analysis of site visits in our previous report revealed that there was a need to further develop mentors' engagement with mentees (Constant et al., 2019). RAND partnered with Sunburst Youth Academy, based in Los Alamitos, California, to design and implement a pilot project to improve the mentoring component. Sunburst and RAND agreed to focus on mentor training and together identified and added modules and exercises to the mentor training session, with the goal of improving mentor communication skills. The recommended training materials on communication skills were adapted from the *Training New Mentors* guide published by the Hamilton Fish Institute on School and Community Violence and the National Mentoring Center, and the additional modules covered two topics—active listening and empathy—and included three activities (Hamilton Fish Institute on School and Community Violence and The National Mentoring Center at Northwest Regional Education Laboratory, 2007). A survey administered to mentors after the training revealed that mentors were positive toward both the overall training and specific communication skills module and related exercises. Follow-up surveys will be administered in the Post-Residential Phase to monitor mentor implementation of these communications skills in their interaction with mentees.

Sunburst and RAND are also implementing a continuation of the pilot by testing the same training module with a second cohort of mentors, as well as requesting that mentors take an online training course titled *Growth Mindset for Mentors*—which was developed by the Project for Education Research that Scales and MENTOR[6]—to help mentors develop growth mindset strategies that can help them to support their mentees. Mentors are expected to complete the training before graduation. A quarterly survey will be administered by Sunburst Youth Academy to track the application of active listening skills and growth mindset strategies and to examine the evolution of the quality of the mentor-mentee relationship over time.

Collecting Outcome Data Through an Alumni Survey

One of the key difficulties reported by ChalleNGe sites is getting accurate placement information from program graduates in the Post-Residential Phase and then tracking their progress on a range of outcomes after they complete the Post-Residential Phase. In response, the RAND research team worked with one of the ChalleNGe sites to develop and implement a short survey to collect information on ChalleNGe alumni experiences and outcomes, including educational attainment, labor market experience, and family formation. We will present the results from this pilot project in a future report.

[6] MENTOR, or the National Mentoring Partnership, is a nonprofit 501c(3) organization that advocates and promotes mentorship for America's youth. For more information, see MENTOR, The National Mentoring Partnership (undated).

Conclusion

Data collected over the past four years indicate that cadets across the ChalleNGe program continue to make progress in many areas. The data also reveal that although all ChalleNGe sites carry out a core foundation of activities, there is considerable site-level variation in this regard. Further research, which the RAND team is undertaking, is needed to better understand how and to what extent this variation is correlated with certain outcomes of interest.

A key shortcoming of this examination of the ChalleNGe program is that information collected to date does not allow the measurement of longer-term outcomes. Developing such measures will continue to be a primary focus of this project, and progress on this front will be documented in future reports. The overarching goal of this project is to help ChalleNGe sites track their progress and inform implementation of program improvements.

Acknowledgments

We are grateful to the staff of the National Guard Youth ChalleNGe program. We thank the central administrative staff who assisted with many aspects of this research and the staff at each location who responded to our data request in a timely fashion and also provided detailed and thoughtful information during the course of our site visits. We are also grateful to our RAND colleagues for their support: Craig Bond and James Powers reviewed various portions of the report, and Emily Payne provided administrative support. Joy Moini, Nicholas Broten, Robert Bozick, Thomas Trail, and Kathryn Edwards made other important contributions to this report. We also thank Matthew Byrd and Amanda Wilson for managing the publications process as well as Samantha Bennett for carefully editing the report. Gabriella Gonzalez of RAND and David DuBois of the University of Illinois at Chicago provided reviews to ensure that our work met RAND's high standards for quality. Jennifer Buck of Spectrum provided the graphic describing the logic model, and we thank Jennifer Demar for developing the programming script to assist ChalleNGe sites with compiling their data. We thank all who contributed to this research or assisted with this report, but we retain full responsibility for the accuracy, objectivity, and analytical integrity of the work presented here.

Abbreviations

ABE	Adult Basic Education
BMI	body mass index
CCR	College and Career Readiness
CCSS	Common Core State Standards
CTE	Career and Technical Education
DRC	Data Recognition Corporation
GED	General Education Development
HiSET	High School Equivalency Test
HSLS	High School Longitudinal Study
P-RAP	Post-Residential Action Plan
RCT	randomized controlled trial
TABE	Tests of Adult Basic Education
TASC	Test Assessing Secondary Completion
TOC	theory of change
WBL	work-based learning

CHAPTER ONE

Introduction: The National Guard Youth ChalleNGe Program

The National Guard Youth ChalleNGe program is a residential, quasi-military program for young people ages 16 to 18 who have left high school without a diploma or are at risk of dropping out because they are unlikely to earn sufficient credits to graduate, given their age and associated grade level. ChalleNGe participants (or *cadets*) may be referred by school counselors or other school officials, law enforcement or the juvenile justice system, or other members of the community. Programs do require, however, that young people who participate do so voluntarily, and parents or guardians must consent to this participation.

Participating states operate the program through their state National Guard organizations with supporting federal funds and oversight. The National Guard is responsible for all day-to-day operational aspects of the program; the Office of the Secretary of Defense provides oversight. States are required by federal law to contribute at least 25 percent of the operating funds. The first ten ChalleNGe sites were established in the mid-1990s; today, there are 39 ChalleNGe sites in 28 states, the District of Columbia, and Puerto Rico. Around 234,000 young people have participated in the ChalleNGe program, and roughly 174,000 have completed the program. Table A.1 in Appendix A includes a list of all ChalleNGe sites.

ChalleNGe's stated mission is "to intervene in and reclaim the lives of 16–18-year-old high school dropouts, producing program graduates with the values, life skills, education, and self-discipline necessary to succeed as productive citizens."[1] ChalleNGe is based on eight core components: leadership and followership, responsible citizenship, service to community, life-coping skills, physical fitness, health and hygiene, job skills, and academic excellence. ChalleNGe's overarching goal is to be recognized as the nation's premier voluntary program for 16–18-year-olds who struggle in a traditional high school setting, serving all U.S. states and territories. Previous research has found that ChalleNGe has a positive influence on participants' near-term labor market outcomes (Bloom, Gardenhire-Crooks, and Mandsager, 2009; Millenky, Bloom, and Dillon, 2010; Millenky et al., 2011) and is cost-effective (Perez-Arce et al., 2012).[2]

---

[1] The mission statement can be found in previous annual reports to Congress (for example, National Guard Youth ChalleNGe, 2015, p. 2) as well as the ChalleNGe website (National Guard Youth ChalleNGe, undated). The mission statement appears to be widely shared across ChalleNGe sites. It is quoted in various materials and briefings used at the sites and was included in briefings that formed part of our site visits.

[2] Researchers at MDRC, an organization that conducts rigorous research on several social policy areas, employed a randomized controlled trial (RCT) to evaluate the effects of ChalleNGe by comparing a treatment group (those who participated in ChalleNGe) with an otherwise similar control group that was not randomly assigned to participate in ChalleNGe. The researchers collected information using a survey at nine months, 21 months, and 36 months following entry into the study (Bloom, Gardenhire-Crooks, and Mandsager, 2009; Millenky, Bloom, and Dillon, 2010; Millenky et al., 2011). The

The RAND Corporation's ongoing analysis of the ChalleNGe program has two primary objectives. First, on an annual basis throughout the project, we collect and analyze data from each site in support of the program's yearly reports to Congress; this is the fourth such RAND report from this project.[3] The first, second, and third reports included information on ChalleNGe classes that began in 2015, 2016, and 2017, respectively, as well as a description of the ChalleNGe logic model and analyses of Tests of Adult Basic Education (TABE) scores, body mass index (BMI), and mentor reporting. All four reports published to date aim to lay the foundation for developing longer-term metrics of cadet success. This report is designed as a stand-alone document; for this reason, it includes some information in common with the previous reports. However, earlier reports (Constant et al., 2019; Wenger, Constant, and Cottrell, 2018; Wenger et al., 2017) include additional analytic details on the TABE; BMI; and classes from 2015, 2016, and 2017.

Our second objective is to develop a rich and detailed set of metrics to capture more information about the long-term effectiveness of the program. To this end, we are undertaking a series of analytic efforts and pilots focusing on various aspects of the ChalleNGe program. In this report, we share preliminary findings from one analytic effort and one pilot, with more-detailed findings to be shared as a separate stand-alone report or in the project's final research report. We provide a synthesis of findings from an analytic effort involving the examination of Career and Technical Education (CTE) practices across program sites. The pilot involves the design and implementation of mentor training that builds on findings from a previously reported-on analytic effort on best-practices in mentoring (Constant et al., 2019). We also describe two other analytic efforts on occupational credentialing and mental health supports for cadets, and another pilot that involves the implementation of an alumni survey undertaken by one of the ChalleNGe sites. A summary of the main project activities is provided in Table 1.1.

The connecting seam across the study's associated analytic efforts is the program logic model. The logic model describes the design of the National Guard Youth ChalleNGe program, the various activities that program sites across states implement, and the expected outputs and outcomes.[4] The data collected for the annual reports include much of the information related to inputs, activities, outputs, and short-term outcomes in the logic model. The analytic efforts and the pilots are intended to address gaps in terms of data collection (particularly those related to long-term outcomes) and to better understand program design and implementation issues, such as improving the mentoring component. The end result of analytic efforts and pilots over the course of the study will be (1) four annual reports; (2) shorter reports summarizing the findings of each analytic study; and (3) a final research report that describes current efforts and suggested strategies to collect long-term outcomes that program sites can use to track progress toward achieving their mission. The detailed findings from the pilots will inform data collection and will be incorporated into the final research report.

RAND study used the MDRC findings to conduct a cost-benefit analysis of the program, factoring in the projected lifetime earnings, given higher educational attainment, from participation in ChalleNGe (Perez-Arce et al., 2012).

[3] This report draws heavily on information provided in our previous reports (Constant et al., 2019; Wenger, Constant, and Cottrell, 2018; Wenger et al., 2017).

[4] We discuss the logic model in more detail in Chapter Three, and it is illustrated in Figure 3.1.

Table 1.1
Main Project Activities and Deliverables

	2016	2017	2018	2019	2020
Project duration	September				June
Site visits					
Data calls to sites					
1st		1st annual report			
2nd			2nd annual report		
3rd				3rd annual report	
4th					4th annual report
Analytic efforts and pilots		Fall 2017 to spring 2020			
Final research report					Final research report

In the remainder of this chapter, we provide additional background information on the ChalleNGe program. We then describe in more detail the focus of this report and the methodologies we used. We conclude with a road map for the remainder of the report.

The ChalleNGe Model

The ChalleNGe program has several unique characteristics. Because cadets generally take part in the program at a site located in the state where they live, not all young people have access to the program. Recruitment for the program varies from site to site. Typically, to ensure broader coverage, program sites conduct regular outreach to high schools (especially to counselors), organizations that run out-of-school programs serving young people, and other community-based health and education organizations that serve underprivileged youth and their families. Program representatives conduct site visits, give presentations about the program, and distribute marketing materials. High school counselors refer students to the program, but, in many cases, students or their parents reach out to the program directly after attending an informational event. Programs also rely on word of mouth from graduates, family members, peers, and high-profile community members who support the program, especially in smaller and tight-knit communities. In some cases, young people are referred to the program by members of the juvenile justice system. Participation is voluntary, and there is no tuition cost to the cadet or his or her family. Cadets must apply to the program, however, and most sites have a "packing list" of items that cadets must bring on their first day. Many sites also require applicants to complete an interview or attend an information session at the site (or both). Most sites do not have minimum standardized test score requirements. Applicants must not be currently awaiting sentencing, on parole, or on probation for anything other than a juvenile offense; also, they must not be under indictment or accused or convicted of a felony (U.S. Department of Defense Instruction 1025.8, 2002).

The ChalleNGe program runs for a total of 17.5 months, broken into a 5.5-month Residential Phase and a 12-month Post-Residential Phase. During the Residential Phase, cadets reside at the program site in a barrack-like atmosphere, wear uniforms, and perform activities generally associated with military training (e.g., marching, drills, physical training). During the first two weeks of the program, referred to as the Pre-Challenge or Acclimation period, cadets learn to adjust to the new environment, as well as to the expectations that the ChalleNGe program requires for success. Coursework begins at the end of the Acclimation period. For the next five months—the main part of the Residential Phase—cadets attend classes for much of the day. The academic curriculum varies across sites; this variation is a result of program history, state context, and choices made by program leadership in each state.[5] Some sites focus on the completion of a GED or High School Equivalency Test (HiSET) credential. At other program sites, cadets have the option to earn high school credits that they can use to transfer to a high school at the end of the ChalleNGe program and go on to earn a high school diploma. Still other ChalleNGe sites award high school diplomas to cadets who complete the state requirements for high school graduation. Some sites give cadets the option to choose among these models.

Not all cadets complete the 5.5-month Residential Phase of the ChalleNGe program (completion is referred to as *graduation*). Most cadets who leave the program prior to graduation choose to withdraw, but sites can and do dismiss cadets who violate key policies. Cadets are not enlisted in the military during the Residential Phase, and there is no requirement of military service following completion of the program.

ChalleNGe places considerable focus on the development of noncognitive or socioemotional skills, such as having positive interpersonal relationships, developing goals and detailed plans to accomplish those goals, anger management, and attention to detail, among others. The program is based on the following eight core components:

- leadership and followership
- responsible citizenship
- service to community
- life-coping skills
- physical fitness
- health and hygiene
- job skills
- academic excellence.

[5] Initially, the academic focus of the ChalleNGe program was on completing a General Education Development (GED) credential. However, some sites have transitioned to awarding credit recovery and high school diplomas. This transition has occurred as a result of several factors. First, leadership at some sites believes strongly that a high school diploma, rather than an equivalency degree, is needed to better ensure success in postsecondary education and the labor market, as well as eligibility for military enlistment. These sites have expended a concerted effort to make this transition. Second, some sites have attained affiliation with a local school district or the state as a charter school. This allows them to award high school diplomas or credits through recovery programs that are recognized by state and local authorities. Finally, sites are also responding to their local communities: In certain state and local contexts, parents and incoming cadets want to pursue a high school diploma rather than an equivalency degree.

Each ChalleNGe site is charged with developing cadets' skills and abilities in all eight areas. Mentorship plays a key role: Each cadet has a mentor, and the relationship between cadet and mentor is intended to continue for at least 12 months after the cadet graduates from the Residential Phase (in other words, through the Post-Residential Phase). The ChalleNGe mentoring model is largely youth-initiated: Cadets are encouraged to nominate their own mentors, and most do.

While there are no formal, professional qualifications for being a mentor, mentors must meet a set of criteria including minimum age. Mentors also must be the same gender as the cadet, be of good standing in the community, generally live in the same community as the cadet, not be an immediate family member of the cadet, and be willing to commit time to training and to attending regular meetings with the cadet. Mentors, who receive in-person training from ChalleNGe staff, are volunteers (i.e., they are not compensated). Mentors also maintain contact with program staff throughout the Post-Residential Phase. If cadets are not able to identify an appropriate mentor, ChalleNGe staff work to recruit one.

The ChalleNGe model has been found to be effective through an RCT; youth who participate go on to complete more postsecondary education than youth who do not participate, and youth who attend the program are more likely to participate in the labor force when compared with similar young people who do not attend the program (Bloom, Gardenhire-Crooks, and Mandsager, 2009; Millenky, Bloom, and Dillon, 2010; Millenky et al., 2011). In a separate and careful analysis of the costs and benefits based on the outcomes from the RCT, RAND researchers found that ChalleNGe is cost-effective, producing approximately $2.66 in benefits (appropriately discounted) for each $1.00 invested (Perez-Arce et al., 2012).[6] The differences observed in the RCT included longer-term outcomes, such as GED attainment, traditional high school diploma attainment, and college attendance, as well as employment and earnings up to three years after graduation; these outcomes are the reason the program was found to be cost-effective. These longer-term outcomes were collected specifically to conduct the RCT; sites do not regularly collect such information from their graduates. In many cases, the outcomes were self-reported in surveys administered as part of the study. The self-reporting could have influenced some outcomes; for example, crime- and health-related outcomes were found to be similar between youth who participated in ChalleNGe and youth who did not. An important limitation of the RCT and the RAND cost-benefit analysis based on the RCT results is that the positive effects of the program on youth were detected using only a subset of ChalleNGe sites. (For a more detailed description of previous research on the ChalleNGe program, see Wenger et al., 2017.)

Focus of This Report and Methodology

This report, the fourth in a series of annual reports produced by RAND for this project, serves two purposes. The first is to *provide a snapshot of the ChalleNGe program on a variety of quantitative indicators during 2018–2019. The second purpose is to develop a set of metrics related to the*

[6] Costs included not only the operating costs of the program but also the opportunity costs of those participating. For more details about the RCT and the differences observed between ChalleNGe participants and similar young people who did not enter ChalleNGe, see Bloom, Gardenhire-Crooks, and Mandsager, 2009; Millenky, Bloom, and Dillon, 2010; and Millenky et al., 2011.

long-term effects that ChalleNGe has on participants after they leave the program; these metrics will help document the extent to which the ChalleNGe program is achieving its mission.

Given the twofold purpose of this report and our larger research agenda regarding the ChalleNGe program, we combine several methodologies: collecting quantitative data from each ChalleNGe site, developing tools to help determine the preferred outcome metrics from the ChalleNGe program, collecting qualitative data through site visits, and planning and beginning to carry out a series of other analytic efforts. We describe each of these efforts here and present additional detail in Chapter Three.

To provide a snapshot of the ChalleNGe program during 2018–2019, we include information gathered from individual ChalleNGe sites in July and August 2019. Much of this program-level information is typical of what was included in the first three annual reports. We collected and reviewed information from each site on program characteristics; 2018 budget and sources of funds; number of applicants, participants, and graduates; credentials awarded; and metrics of physical fitness and community service or engagement. We also collected information on staffing, the dates classes began and ended, and postresidential placements. We requested and received the information through secure data transfers (although we requested no identifying information). We specified that sites should include information from the two classes that began in 2018 (most sites start classes in January and July, but some sites run on different schedules). This information meets the program's current annual reporting requirements and will be used in the program's 2019 report to Congress.[7] In Chapter Two, we provide program- and class-specific data and some analysis of this information across programs.

As part of our data collection, we also requested cadet-level information on graduation, credentials awarded, changes in TABE grade-equivalent scores, and placements during the Post-Residential Phase.[8] Annual reports for the ChalleNGe program published prior to 2017 included only site-level measures and metrics,[9] such as the average gain in TABE grade-equivalent scores or the number of cadets placed; they do not include any cadet-level information.[10] Achieving *key levels* on the TABE predicts other relevant outcomes, such as passing the GED exam, although, as we will note in the next chapter, the mapping of the new version of the TABE (11/12) to outcomes of interest is not currently available. Where we can, we use the cadet-level information to report a series of metrics based on achieving key TABE levels (we developed these levels during previous years). We include some analyses of this informa-

---

[7] See 32 U.S.C. §509(k) for annual reporting requirements.

[8] TABE is currently developed by DRC | CTB, and its suite of tests is specifically designed to assess the basic skills of adult learners. According to the TABE website, workforce development programs in most U.S. states, whether funded or not funded by the federal Workforce Innovation and Opportunity Act, use TABE to assess the basic skills of individuals participating in their programs (TABE, undated). All ChalleNGe programs administer TABE to cadets at the beginning of the program and prior to graduation to measure academic achievement in math and reading and to maintain a key metric by which to track cadet learning progress. TABE results are reported in past analyses; see, for example, the 2015 annual report (National Guard Youth ChalleNGe, 2015).

[9] In technical terms, a *metric* is a specific value, while a *measure* refers to an activity, output, or outcome (National Research Council, 2011). Thus, the number of cadets who graduate from ChalleNGe could be considered a measure, while an overall cadet graduation rate of 80 percent could be considered a metric. In this report, we more frequently refer to a *metric* to imply a specific measure, which may eventually have a goal associated with it.

[10] *Average gain* in TABE grade-equivalent scores is widely used but problematic (see Lindholm-Leary and Hargett, 2006, as well as Wenger et al., 2017).

tion in Chapter Two; see Wenger, Constant, and Cottrell (2018) and Wenger et al. (2017) for additional details.

To develop a set of metrics to gauge the longer-term outcomes of program graduates, we developed a ChalleNGe-specific logic model for defining the longer-term outcomes. We began by developing two tools: a theory of change (TOC) and a program logic model. The TOC and the logic model serve as operational tools to guide the development of metrics, monitor progress toward achieving the program's central goals, and evaluate its effectiveness. The TOC is a conceptualization of the mechanisms by which solutions can be developed to address a complex social problem; a program logic model delineates the inputs, processes or activities, expected outputs, and desired outcomes of a specific program designed to address a problem (Shakman and Rodriguez, 2015). A logic model builds on a TOC and includes more information to develop metrics or indicators to monitor progress in implementation. To this end, the ChalleNGe logic model includes a detailed list of longer-term outcomes that we might expect to see in ChalleNGe graduates and that might ultimately form the basis of evaluating the program's effectiveness.[11] We included detailed information about these tools in our earlier reports (Wenger et al., 2017; Wenger, Constant, and Cottrell, 2018; Constant et al., 2019), but we also include the logic model in Chapter Three of this report.

Next, we developed a detailed site-visit protocol (based partly on the logic model) and set up a schedule that allowed us to visit each ChalleNGe site over the course of this project. In 2019, we completed the remaining nine visits to program sites, thus conducting at least one visit to each ChalleNGe site since the commencement of the study in 2016.[12] During each visit, we interviewed program leadership and staff, and we collected detailed information on the program and any specific initiatives undertaken at a particular site. Among the topics we covered at each visit were the site's mission and general approach, practices used to recruit potential cadets, training of mentors, instructional practices, information about occupational training offered to cadets, placement strategies, data collection strategies, and disciplinary policies (see Table 1.2 for a list of staff we typically interviewed and topics covered). Since our focus is on developing metrics that capture long-term outcomes, we also asked staff for their input on the types of metrics they would like to learn more about. As noted in our third report, staff are interested in learning about graduates' postsecondary education pursuits, employment, and indicators of health and well-being and successful transitions into adulthood.

The RAND research team will draw on the data collected from sites to extract findings that are relevant to identifying metrics for measuring long-term outcomes. This topic will be the focus of the project's final report, to be published in summer 2020. The site visits were particularly helpful in identifying the opportunities for and challenges to collecting long-term metrics from a site perspective, as well as understanding the state and local policy context and the implications of program access to, and use and reporting of, postresidential data.

In addition to site visits and our ongoing analytic work, the study team identified opportunities to support program sites that are conducting pilot projects. These pilots are run by the selected program sites with RAND team advice. The RAND study team helped pilot sites develop tools to assess the implementation and results of the pilots. Currently, these pilots include approaches to collecting data from graduates beyond the one-year Post-Residential

[11] For more information on logic models, see (among others) Knowlton and Phillips (2009), as well as Shakman and Rodriguez (2015).

[12] In some cases, we conducted more–limited-in-scope follow-up visits to certain sites.

Table 1.2
Topics Covered in the Site Visit Protocol and Main Sources of Information

Staff Interviews	Topics Covered
Director and deputy director	Program mission; relationship to community and parents/ guardians; staff hiring, recruitment, and performance appraisal; cadet outcomes; finance and resources; desired indicators
Recruiting, placement, and mentoring staff	Recruiting; mentor assignment and training; postresidential preparation, placement, and tracking; desired indicators
Commandant (head of the cadre)	Cadet discipline, cadet schedule, barracks and general environment, desired indicators
Lead instructor and instructional staff	Curriculum and instruction; Career and Technical Education, occupational training; desired indicators
Management information services lead	Data management

Phase and improving mentor training. The RAND study team will help sites document the results of pilot implementation and will make the findings available to all sites. We describe our analytic plans in more detail in Chapter Three.

Organization of This Report

The remainder of this report consists of three chapters and two appendixes:

- Chapter Two provides a snapshot of the ChalleNGe program in 2018–2019. This snapshot includes information about recent classes, which is comparable with information in past reports, as well as information on the proportion of cadets meeting key TABE levels, cadets' contributions to their communities, placement rates after cadets leave the program, details about a few key aspects of each program, and analyses of trends over time.
- Chapter Three discusses the logic model and the importance of measuring the longer-term outcomes of the program. This chapter also describes various analyses in support of measuring longer-term outcomes, as well as two pilot studies.
- Chapter Four presents concluding thoughts.
- Appendix A includes a complete list of the ChalleNGe programs and detailed information collected from each program.
- Appendix B includes information that explains the transition to the new version of the TABE (11/12).

CHAPTER TWO

Data and Analyses: 2018 ChalleNGe Classes

In this chapter, we analyze and present information on many aspects of the ChalleNGe program. The focus in the first section of the chapter is on cross-site data; we include measures of the overall numbers of participants, as well as credentials awarded, standardized test scores, and measures of citizenship, community service, and physical fitness. These measures span many of the core components of ChalleNGe. In later sections of the chapter, we discuss some recent changes to the TABE (the primary standardized test used by all ChalleNGe sites), but we also present information on staffing, as well as data on cadet placements during the Post-Residential Phase of ChalleNGe.

As in the past three years, in this report, we base our analyses on quantitative information received from each site. When we collected the data for this report (July–August 2019), 39 ChalleNGe sites were operating; each site provided data for this report.[1] Therefore, our sample differs slightly from the samples used in earlier reports. Because new sites tend to be relatively small and because the vast majority of sites have remained consistent across reports, differences due to site-level changes are likely to be minor. For example, a little more than 1 percent of cadets from classes 50 and 51 attended ChalleNGe in Tennessee; this site was not yet able to report data in 2016. About 5 percent of cadets in this report attended new sites in California or Georgia; these sites were not operational in 2016, but these cadets would likely have had access to ChalleNGe through other sites established earlier in those states. Therefore, despite the site-level changes, we expect that most results are comparable across reports.

As in past years, we began the data collection process by developing and sending a spreadsheet template to each site to record program-level and cadet-level information; a staff member (the primary data point of contact) was responsible for gathering the information at the program and cadet levels. This template, first developed in 2016, was revised in 2017, 2018, and then again in 2019 to reflect lessons learned from each round of data collection, as well as to meet evolving information needs. The RAND study team implemented improvements and refinements to the indicators and added questions informed by site visits. However, many of the fundamental data elements collected have remained the same across the years. The spreadsheet template was distributed to all sites during each of the four rounds of data collection.

[1] There have been changes in the number of ChalleNGe sites operating over time. In past years, two separate ChalleNGe sites operated in Texas; at this point, the two sites have been combined into a single site. Therefore, we report data for only a single Texas site. Three sites (Georgia-Milledgeville, Tennessee, and California-Discovery) were not yet operational when we began tracking each site in 2016–2017. During our 2017 data collection, the Puerto Rico ChalleNGe site was not operational due to damage from Hurricane Maria. For these reasons, the total number of ChalleNGe sites varies across our reports.

Although we requested no identifying information, data were transmitted to RAND researchers through a secure file transfer protocol link. To facilitate this process, we contracted with the organization that built the database used by many of the sites; this organization provided information to the sites to assist them in extracting the required data elements. We continue to refine this data collection process with the joint goals of lowering the burden on individual ChalleNGe sites and further improving the accuracy of the information used for ChalleNGe annual reports.

To ensure data fidelity, we implemented several key procedures as part of our quality assurance process, including

- confirming with sites when sections of the data they reported were incomplete or missing
- exploring outliers
- comparing counts and averages across sites and classes
- comparing trends by site and classes against previous ChalleNGe reports
- comparing site data with program-wide data to ensure broad consistency.

Despite our data quality assurance efforts, it is important to recognize that there are likely to be at least some errors in the data due to the nature of the data collection method; we suspect that errors are especially likely to occur in the graduate placement information. During our site visits, program administrators frequently acknowledged the difficulty of obtaining and verifying placement data from graduates and their mentors, and the further out after graduation, the more difficult it becomes. Thus, while placement data provide general information about what graduates of ChalleNGe are doing in the Post-Residential Phase, it is important to keep in mind that the placement data lack enough detail to be able to ascertain the quality of the placement status. For example, ChalleNGe sites collect information from mentors on whether a graduate is employed, but the programs do not consistently collect materials to verify employment or to determine the number of hours worked and wages earned. Thus, there is likely to be a wide range in the quality of the placement information that programs collect and report on. We are currently developing a pilot program to test several ideas that have the potential to increase reporting by mentors. A description of this pilot is provided in Chapter Three, and we will more fully document the results of this pilot program in a future report.

We begin by presenting a summary of the information from all reporting sites. These metrics serve to measure the overall progress of the ChalleNGe program in terms of the number of young people who participated in ChalleNGe in 2018 (these classes are referred to by ChalleNGe staff as Classes 50 and 51). We also include tallies of the total number of academic credentials awarded, the hours and value of community service documented, and the overall placement rates.[2] We then present this information in a less-aggregated manner, for each site and by class. In the next section of this chapter, we present a detailed analysis of the data on cadets' TABE scores and use RAND-developed metrics to show the number of cadets who achieved key TABE milestones. In that section, we also discuss the recent changes to the TABE test and the implications for the ChalleNGe program. We then present some information on

---

[2] In some cases, we requested similar information at the site and cadet levels; for example, we requested the overall number of credentials awarded, as well as indicators of which cadets received credentials. We found occasional minor discrepancies in these data. When such discrepancies occurred, we reported the numbers calculated from the cadet-level data.

site-level differences in staffing, staff turnover, and starting salaries. We finish by presenting some time trends using information from current and previous data collections.

Cross-Site Metrics for the 2018 Classes

Below, we present summary information on the cadets who entered a ChalleNGe program in 2018. ChalleNGe received 19,257 applications for the classes that began during calendar year 2018. Based on site-specific enrollment criteria, 12,844 young people were accepted by, and chose to enroll in, a ChalleNGe program. Of the 12,844 cadets who enrolled, 9,351 (73 percent) *graduated* from the 5.5-month Residential Phase of ChalleNGe. Most ChalleNGe sites operate on a January-to-June and July-to-December schedule, although a small number of programs operate on different schedules during the year. Thus, we define *2018 participants* as those who attended a ChalleNGe program that started in 2018. In some cases, cadets may have applied in 2017 (e.g., to enter a program that began in January 2018). In most cases, cadets graduated during 2018, but a few programs spanned the 2018–2019 calendar years. Table 2.1 provides a summary of several key ChalleNGe statistics, across all sites.

Figures 2.1–2.7 and Tables 2.2–2.9 include data on several of the core components of ChalleNGe, presented for each site and each class. Detailed tables for the figures are shown in Appendix A.[3] These figures and tables provide a detailed sense of each site's progress on multiple metrics. In some cases, individual data items are noted as missing in this chapter and in Appendix A. When this occurs, we note the specific reason. Some sites failed to report spe-

Table 2.1
ChalleNGe Statistics, 1993–2018

Challenge Statistics	1993–2017	2018[a]	1993–2018
Applicants	389,461	19,257	408,718
Enrollees	221,661	12,844	234,505
Graduates	164,998	9,351	174,349
Academic credentials[b]	100,683	4,092	104,775
Service hours to communities	11,108,561	569,151	11,677,713
Value of service hours	$224,308,850	$14,046,660	$238,355,510

[a] Information in this table was reported by the sites in July and August 2019 and covers Classes 50 and 51; these classes began in 2018. *Applicants* includes all who completed an application (although sites may define application completion in slightly different ways).

[b] *Academic credentials* reflect cadets who graduated and received either a GED or a HiSET or a Test Assessing Secondary Completion (TASC) credential or a high school diploma (limited to one credential per cadet). When we also consider high school credits, over 60 percent of cadets received an academic credential (see Figures 2.2 and 2.3, as well as Table A.3 for more information). Tennessee Classes 50 and 51 did not report credentials; see Table 2.3 for more information. Additionally, programs may have reported the *total* number of academic credentials for earlier classes rather than limiting credentials to one per cadet; therefore, the numbers here and in Table 2.3 may not be comparable with those documented in reports pertaining to ChalleNGe classes that graduated prior to 2015.

[3] Tables A.2–A.4 in Appendix A provide more-detailed information on each ChalleNGe site, including information on staffing, funding, dates when classes began and ended, as well as measures of physical fitness, responsible citizenship, service to community, and detailed placement information on ChalleNGe graduates.

cific data elements, but in other cases, information was not yet available; for example, Class 51 graduates left most programs less than a year prior to our data collection, so no 12-month placement data on these cadets was requested or reported.

All subsequent figures and tables in this chapter include information for each site and class. Full names, locations, and abbreviations for the sites can be found in Table A.1 in Appendix A. The figures in this chapter have accompanying tables with more-detailed data, in some cases, that are provided in the appendix. TABE scores calculated in Figures 2.1–2.7 and Tables 2.2–2.9 include only cadets who graduated from ChalleNGe. The figures and tables are organized by the following metrics:

- **Graduation rate by site (Figure 2.1).** The number of cadets who graduate as a share of those who enter ChalleNGe is a key metric for sites (Figure 2.1). Also important is the number of applicants and graduates, including the targeted number of graduates, which is shown in the appendix (Table A.2). The targeted number of graduates is a key metric for ChalleNGe sites because it is considered in setting their budgets. In the final section of this chapter, we analyze trends in target numbers, applications, and graduates over time.
- **Credentials awarded (Figures 2.2 and 2.3).** Figures 2.2 and 2.3 include the share of credentials attained by graduates for each site for Class 50 and 51, respectively. Table A.3 in the appendix provides the number of graduates by credential. To better determine the proportion of graduates who received at least one credential, we requested that sites report only a single credential for each graduate. Therefore, graduates who received high school credits and a high school diploma are listed as having received only a diploma; those who received a GED or HiSET certification and high school credits are listed as receiving only high school credits (this second case is quite rare). As noted above, a few sites reported these data in an inconsistent manner. Figure 2.1 includes only the more-restrictive definition of *credentials*—passing a standardized test or receiving a high school diploma. Based on this definition, more than 40 percent of ChalleNGe graduates received a credential. But if we also include high school credits as credentials, more than 60 percent of graduates received at least one credential. Figures 2.2 and 2.3 and Table A.3 in the appendix include this broader set of credentials.
- **TABE scores (Tables 2.2–2.7).** We collected information on the total TABE battery, but also on three specific subtests: math, language, and reading. We report additional information on TABE scores for all cadets in a later subsection of this chapter. TABE scores are reported as grade equivalents; for example, a score of 7.5 indicates that the test-taker performed similarly to a typical student at the fifth month of seventh grade. Cadets generally achieved higher TABE scores at the end of ChalleNGe than at the beginning, across sites.
- **Responsible citizenship (Tables 2.8 and 2.9).** Metrics of responsible citizenship include registration for voting (all cadets) and registration for the Selective Service (male cadets). The majority of sites registered 100 percent of eligible cadets for voting and Selective Service.
- **Community service (Figures 2.4 and 2.5).** We report the average hours of community service per cadet, as well as the value of that service. The value of community service is calculated using published figures at the state level for 2015, which are available online (Independent Sector, undated). The value of community service was calculated in the same manner in the previous four annual reports (Constant et al., 2019; Wenger, Constant, and Cottrell, 2018; Wenger et al., 2017; National Guard Youth ChalleNGe, 2015).

Figure 2.1
Graduation Rate, by Site

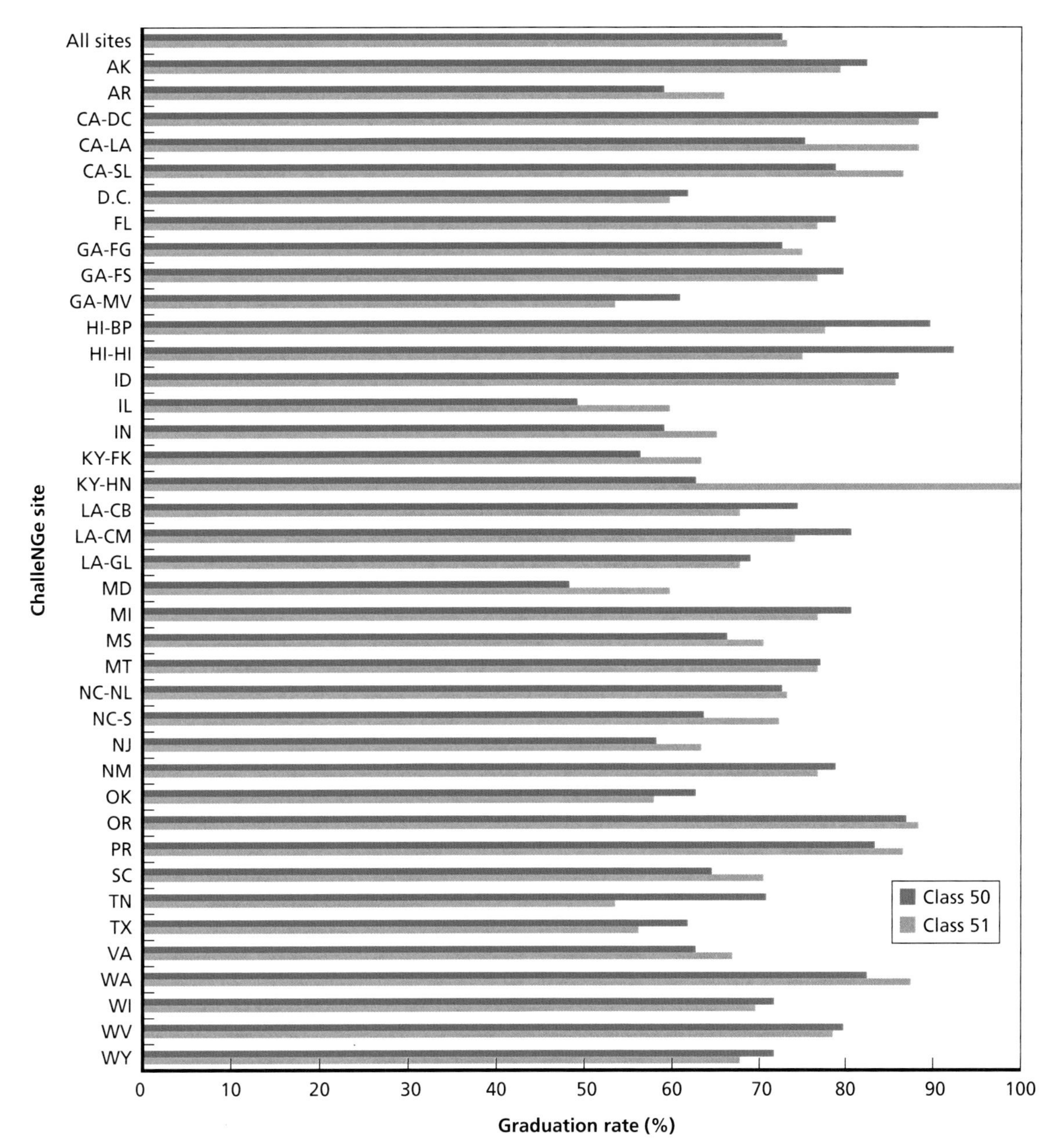

NOTES: Information in this figure was reported by the sites in July and August 2019 and covers Classes 50 and 51, which began in 2018. Graduation rate was calculated as the share of graduates to ChalleNGe entrants. For a detailed table of target, applicants, entrants, and graduates for each program, see Table A.2 in Appendix A. AK = Alaska; AR = Arkansas; CA-DC = California-Discovery; CA-LA = California-Sunburst; CA-SL = California-Grizzly; D.C. = Washington, D.C.; FL = Florida; GA-FG = Georgia–Fort Gordon; GA-FS = Georgia–Fort Stewart; GA-MV = Georgia-Milledgeville; HI-BP = Hawaii–Barbers Point; HI-HI = Hawaii-Hilo; ID = Idaho; IL = Illinois; IN = Indiana; KY-FK = Kentucky-Bluegrass; KY-HN = Kentucky-Appalachian; LA-CB = Louisiana–Camp Beauregard; LA-CM = Louisiana–Camp Minden; LA-GL = Louisiana–Gillis Long; MD = Maryland; MI = Michigan; MS = Mississippi; MT = Montana; NC-NL = North Carolina–New London; NC-S = North Carolina–Salemburg; NJ = New Jersey; NM = New Mexico; OK = Oklahoma; OR = Oregon; PR = Puerto Rico; SC = South Carolina; TN = Tennessee; TX = Texas; VA = Virginia; WA = Washington; WI = Wisconsin; WV = West Virginia; WY = Wyoming.

Figure 2.2
Percentage of ChalleNGe Graduates by Type of Credential Awarded, by Site (Class 50)

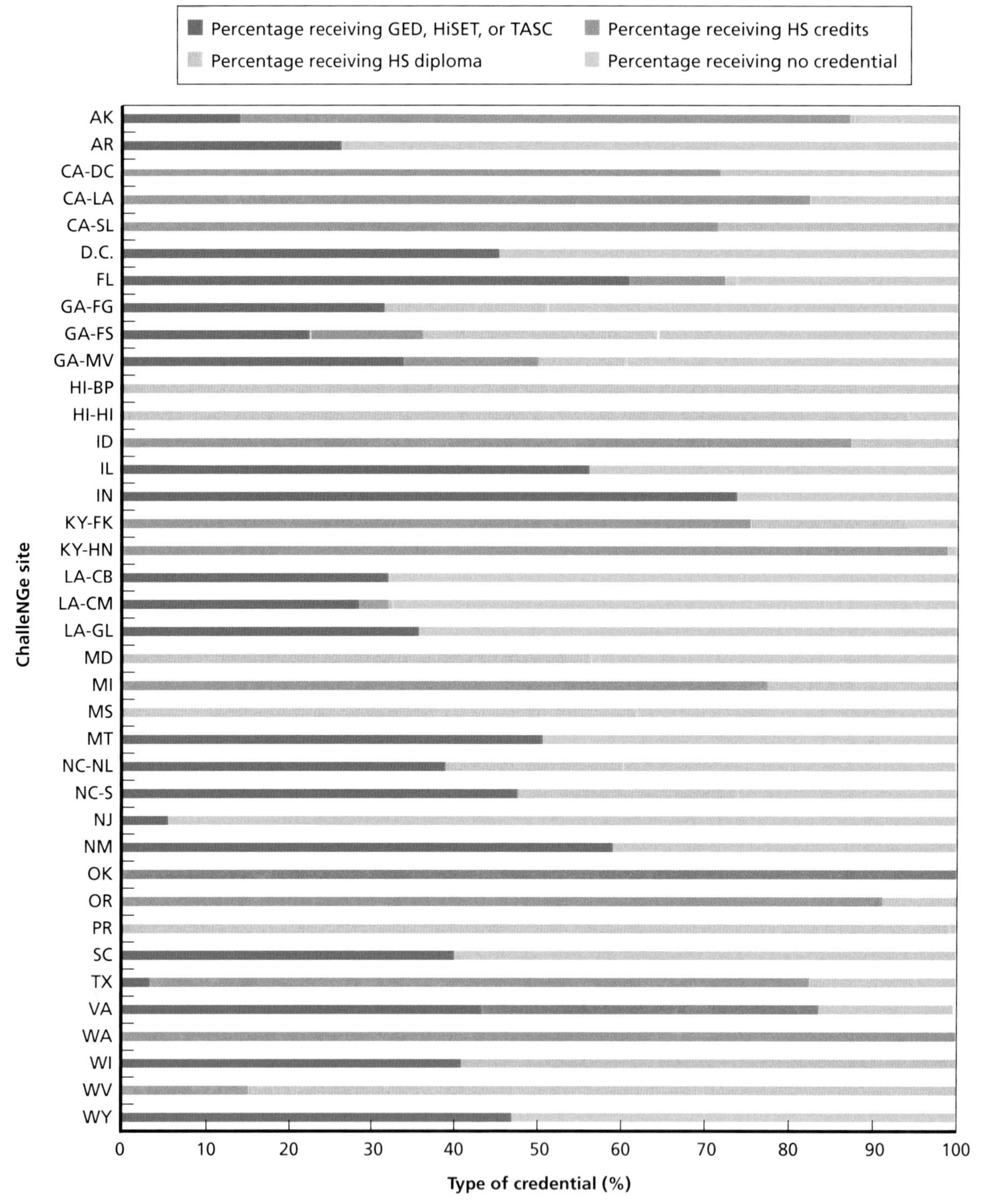

NOTES: Information in this figure was reported by the sites in July and August 2019 and covers Class 50, which began in 2018. Each bar represents the share of credential achieved by graduates of ChalleNGe. Graduates who did not complete a credential fall in the "No credential" category. For a detailed table of the number of graduates by credential for Class 50, see Table A.3 in the appendix. HS = high school. Tennessee did not report.

Figure 2.3
Percentage of ChalleNGe Graduates by Type of Credential Awarded, by Site (Class 51)

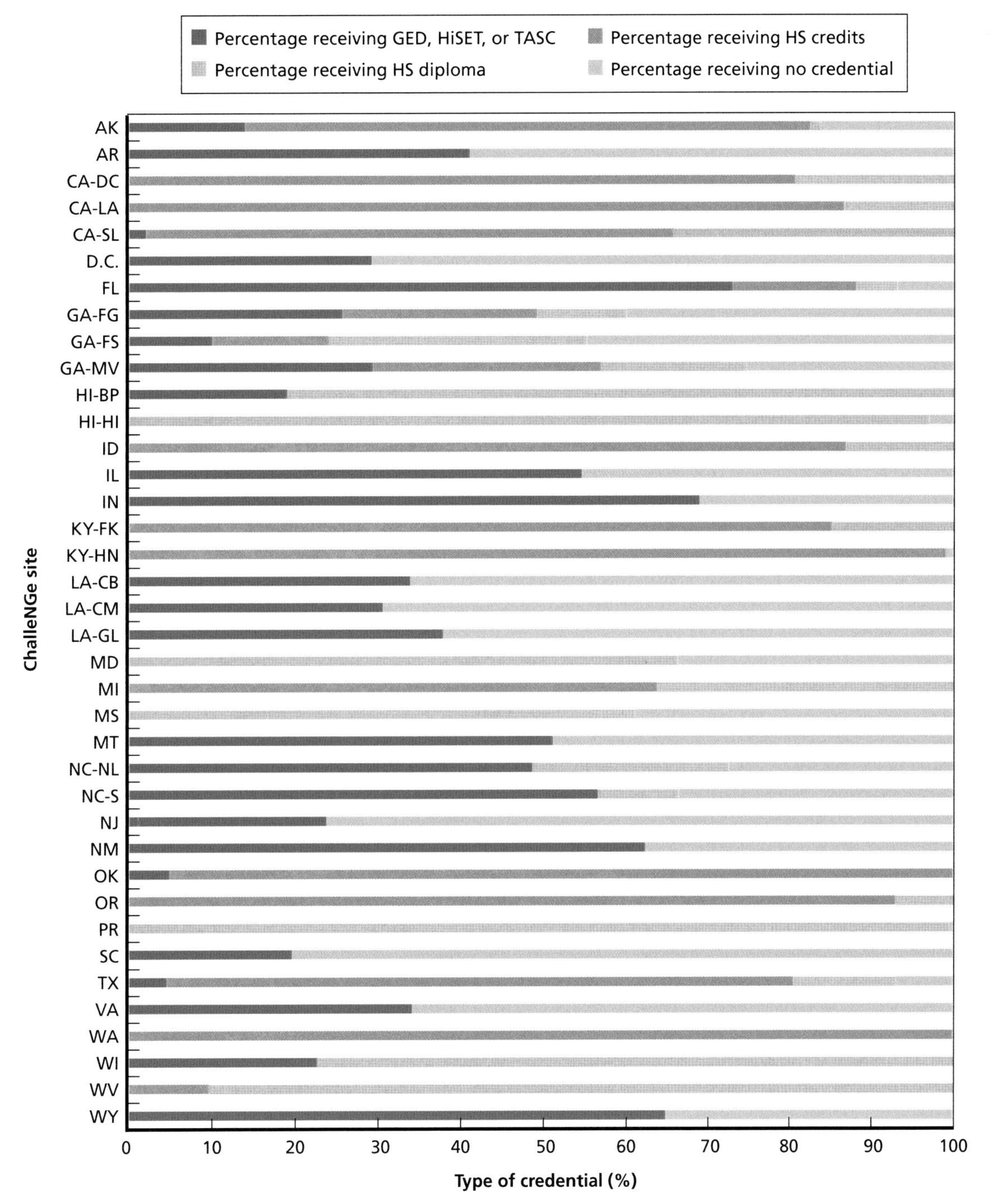

NOTES: Information in this figure was reported by the sites in July and August 2019 and covers Class 51, which began in 2018. Each bar represents the share of credential achieved by graduates of ChalleNGe. Graduates who did not complete a credential fall in the "No credential" category. For a detailed table of the number of graduates by credential for Class 51, see Table A.3 in the appendix. Tennessee did not report.

Table 2.2
Percentage of ChalleNGe Graduates in Pre- and Post-TABE Math Grade-Equivalent, by Site (Class 50)

	Pre-TABE (%)			Post-TABE (%)		
Site	Elementary (Grades 1–5)	Middle School (Grades 6–8)	High School (Grades 9–12)	Elementary (Grades 1–5)	Middle School (Grades 6–8)	High School (Grades 9–12)
All sites	54	25	21	26	31	43
AK	47	22	31	21	26	53
AR	58	23	20	26	47	26
CA-DC	68	23	9	31	41	29
CA-LA	71	13	16	22	32	46
CA-SL	59	19	22	41	32	27
D.C.	48	39	13	26	39	35
FL	28	38	34	7	28	65
GA-FG	68	20	12	47	32	21
GA-FS	46	29	24	1	22	77
GA-MV	49	28	22	25	39	36
HI-BP	47	31	22	19	32	49
HI-HI	72	19	9	69	24	7
ID	40	30	31	7	33	59
IL	51	32	17	18	34	48
IN	44	25	31	18	24	58
KY-FK	75	16	10	32	35	32
KY-HN	88	11	1	19	35	46
LA-CB	47	27	25	15	26	59
LA-CM	58	25	17	21	29	50
LA-GL	40	31	29	13	28	59
MD	68	18	14	31	34	35
MI	65	18	17	50	23	28
MS	48	28	24	5	28	67
MT	44	27	29	28	30	42
NC-NL	53	27	19	53	19	29
NC-S	57	26	17	37	26	37
NJ	31	34	35	15	23	62
NM	52	29	19	16	38	46
OK	45	27	28	31	30	39
OR	52	21	27	20	34	46

Table 2.2—Continued

Site	Pre-TABE (%)			Post-TABE (%)		
	Elementary (Grades 1–5)	Middle School (Grades 6–8)	High School (Grades 9–12)	Elementary (Grades 1–5)	Middle School (Grades 6–8)	High School (Grades 9–12)
PR	91	7	1	40	46	14
SC	62	25	13	41	34	24
TN	43	24	32	*	*	*
TX	49	35	16	48	50	2
VA	69	19	12	51	27	22
WA	58	33	9	20	38	42
WI	35	32	33	23	25	52
WV	50	29	21	31	38	31
WY	36	19	44	26	22	52

NOTES: Information in this table was reported by the sites in July and August 2019 and covers Class 50. Some numbers do not sum to 100 percent due to rounding.

* = did not report.

Table 2.3
Percentage of ChalleNGe Graduates in Pre- and Post-TABE Math Grade-Equivalent, by Site (Class 51)

Site	Pre-TABE (%)			Post-TABE (%)		
	Elementary (Grades 1–5)	Middle School (Grades 6–8)	High School (Grades 9–12)	Elementary (Grades 1–5)	Middle School (Grades 6–8)	High School (Grades 9–12)
All sites 9/10	55	26	19	26	32	42
All sites 11/12	67	29	4	51	37	12
AK[a]	69	26	5	45	38	17
AR[a]	53	42	5	48	47	6
CA-DC	67	23	9	34	25	41
CA-LA	75	16	9	21	24	55
CA-SL	31	33	36	42	37	21
D.C.	66	24	10	39	34	27
FL	35	39	26	10	29	61
GA-FG	69	21	10	35	40	25
GA-FS	39	39	22	3	21	77
GA-MV	39	41	20	27	38	34
HI-BP	58	22	20	25	37	38
HI-HI	63	21	16	51	33	16

Table 2.3—Continued

	Pre-TABE (%)			Post-TABE (%)		
Site	Elementary (Grades 1–5)	Middle School (Grades 6–8)	High School (Grades 9–12)	Elementary (Grades 1–5)	Middle School (Grades 6–8)	High School (Grades 9–12)
ID	36	33	32	10	29	61
IL	58	24	18	19	36	45
IN[a]	72	25	3	72	28	0
KY-FK	51	37	12	37	24	39
KY-HN	81	11	7	37	33	30
LA-CB	58	28	14	21	43	36
LA-CM	52	30	18	13	36	51
LA-GL	50	28	22	22	29	49
MD	61	21	17	20	34	46
MI	58	29	14	37	25	38
MS	44	30	26	8	22	70
MT	49	23	28	26	28	47
NC-NL	44	32	24	33	32	35
NC-S	54	26	20	33	28	39
NJ[a]	58	37	5	44	33	23
NM[a]	79	21	0	64	32	5
OK	49	27	25	35	29	36
OR	51	22	27	23	35	42
PR	89	11	0	46	41	14
SC	68	14	18	30	38	31
TN	*	*	*	*	*	*
TX	46	36	18	35	45	20
VA	62	16	22	45	26	29
WA	61	26	13	15	41	43
WI[a]	68	25	7	51	36	13
WV[a]	69	28	2	51	36	12
WY	32	24	44	21	26	53

NOTES: Information in this table was reported by the sites in July and August 2019 and covers Class 51. Some numbers do not sum to 100 percent due to rounding.

[a] These sites (AK, AR, IN, NJ, NM, WI, and WV) used TABE Survey Form 11/12, while the remaining sites used TABE Survey Form 9/10.

* = did not report.

Table 2.4
Percentage of ChalleNGe Graduates in Pre- and Post-TABE Reading Grade-Equivalent, by Site (Class 50)

	Pre-TABE (%)			Post-TABE (%)		
Site	Elementary (Grades 1–5)	Middle School (Grades 6–8)	High School (Grades 9–12)	Elementary (Grades 1–5)	Middle School (Grades 6–8)	High School (Grades 9–12)
All sites	37	28	36	19	28	54
AK	28	29	42	26	23	51
AR	25	29	46	21	33	46
CA-DC	33	24	43	15	24	61
CA-LA	59	18	23	10	33	58
CA-SL	37	26	36	14	31	55
D.C.	19	48	32	13	29	58
FL	21	34	45	7	24	69
GA-FG	59	23	18	44	28	28
GA-FS	40	34	26	4	20	76
GA-MV	41	33	26	31	26	44
HI-BP	38	30	32	17	28	55
HI-HI	53	29	18	46	37	18
ID	22	28	50	5	26	68
IL	31	30	39	26	33	41
IN	29	33	38	18	32	50
KY-FK	52	24	24	42	29	29
KY-HN	50	25	25	30	42	28
LA-CB	24	32	44	13	23	64
LA-CM	34	28	38	16	26	58
LA-GL	22	31	47	13	22	66
MD	41	31	28	14	37	49
MI	53	24	23	31	30	38
MS	31	26	44	6	24	70
MT	29	25	46	12	38	51
NC-NL	34	32	35	22	33	45
NC-S	30	37	33	10	28	62
NJ	20	23	57	10	36	55
NM	35	34	31	18	27	55
OK	40	22	38	21	27	51

Table 2.4—Continued

Site	Pre-TABE (%)			Post-TABE (%)		
	Elementary (Grades 1–5)	Middle School (Grades 6–8)	High School (Grades 9–12)	Elementary (Grades 1–5)	Middle School (Grades 6–8)	High School (Grades 9–12)
OR	19	24	57	13	27	60
PR	75	11	14	30	8	61
SC	33	40	27	19	31	50
TN	20	29	51	*	*	*
TX	29	29	43	18	34	48
VA	58	20	22	31	35	35
WA	59	20	20	21	33	46
WI	32	30	38	17	33	50
WV	18	38	44	23	34	43
WY	31	23	46	21	26	53

NOTES: Information in this table was reported by the sites in July and August 2019 and covers Class 50. Some numbers do not sum to 100 percent due to rounding.

* = did not report.

Table 2.5
Percentage of ChalleNGe Graduates in Pre- and Post-TABE Reading Grade-Equivalent, by Site (Class 51)

Site	Pre-TABE (%)			Post-TABE (%)		
	Elementary (Grades 1–5)	Middle School (Grades 6–8)	High School (Grades 9–12)	Elementary (Grades 1–5)	Middle School (Grades 6–8)	High School (Grades 9–12)
All sites 9/10	39	27	34	18	28	54
All sites 11/12	58	27	14	45	31	24
AK[a]	57	27	16	36	35	29
AR[a]	54	33	13	43	38	19
CA-DC	40	29	31	23	26	51
CA-LA	71	20	9	8	32	60
CA-SL	38	32	30	9	39	52
D.C.	44	34	22	24	41	34
FL	33	25	42	5	27	68
GA-FG	51	22	27	33	36	31
GA-FS	25	38	37	4	19	76
GA-MV	36	25	38	29	34	36
HI-BP	38	30	32	25	29	46
HI-HI	30	37	33	40	21	39

Table 2.5—Continued

Site	Pre-TABE (%)			Post-TABE (%)		
	Elementary (Grades 1–5)	Middle School (Grades 6–8)	High School (Grades 9–12)	Elementary (Grades 1–5)	Middle School (Grades 6–8)	High School (Grades 9–12)
ID	22	23	55	7	21	72
IL	33	40	27	17	36	48
IN[a]	77	15	8	74	23	2
KY-FK	37	22	41	51	32	17
KY-HN	80	13	7	44	26	30
LA-CB	50	24	26	15	38	46
LA-CM	33	29	38	13	28	59
LA-GL	31	32	37	20	24	56
MD	39	31	31	12	21	66
MI	49	23	28	32	28	40
MS	31	26	43	11	19	70
MT	25	25	50	17	24	59
NC-NL	35	29	36	24	32	44
NC-S	25	45	30	11	30	59
NJ[a]	54	29	16	32	30	38
NM[a]	62	27	11	54	25	20
OK	36	21	43	25	23	52
OR	18	30	52	10	28	62
PR	73	5	22	24	11	64
SC	33	38	30	29	36	35
TN	*	*	*	*	*	*
TX	18	38	45	21	35	44
VA	40	21	40	21	32	47
WA	33	32	35	10	34	55
WI[a]	53	25	22	46	29	25
WV[a]	58	30	12	49	28	23
WY	15	35	50	9	26	65

NOTES: Information in this table was reported by the sites in July and August 2019 and covers Class 51. Some numbers do not sum to 100 percent due to rounding.

* = did not report.

[a] These sites (AK, AR, IN, NJ, NM, WI, and WV) used TABE Survey Form 11/12, while the remaining sites used TABE Survey Form 9/10.

Table 2.6
Percentage of ChalleNGe Graduates in Pre- and Post-TABE Total Battery Grade-Equivalent, by Site (Class 50)

	Pre-TABE (%)			Post-TABE (%)		
Site	Elementary (Grades 1–5)	Middle School (Grades 6–8)	High School (Grades 9–12)	Elementary (Grades 1–5)	Middle School (Grades 6–8)	High School (Grades 9–12)
All sites	49	26	25	22	26	52
AK	36	31	33	22	27	50
AR	40	34	26	22	41	37
CA-DC	*	*	*	*	*	*
CA-LA	62	21	16	19	23	58
CA-SL	46	30	24	20	29	52
D.C.	39	32	29	16	39	45
FL	28	31	40	6	25	69
GA-FG	72	16	13	49	27	24
GA-FS	52	26	22	1	24	75
GA-MV	49	27	24	26	41	33
HI-BP	48	24	28	24	21	54
HI-HI	70	19	10	69	24	7
ID	30	27	43	7	22	71
IL	43	34	23	28	31	41
IN	39	29	32	21	26	53
KY-FK	76	11	13	47	29	24
KY-HN	86	6	8	42	26	32
LA-CB	36	31	33	11	26	63
LA-CM	50	25	25	19	26	55
LA-GL	30	30	41	14	17	69
MD	63	20	17	27	31	42
MI	71	13	16	53	18	29
MS	30	39	31	5	17	78
MT	40	28	32	19	35	46
NC-NL	54	28	18	39	33	28
NC-S	49	38	14	26	33	41
NJ	30	30	41	15	29	56
NM	51	31	18	24	24	52
OK	50	19	30	30	27	43

Table 2.6—Continued

Site	Pre-TABE (%)			Post-TABE (%)		
	Elementary (Grades 1–5)	Middle School (Grades 6–8)	High School (Grades 9–12)	Elementary (Grades 1–5)	Middle School (Grades 6–8)	High School (Grades 9–12)
OR	38	28	34	18	28	54
PR	81	14	5	0	0	100
SC	47	34	19	26	38	36
TN	*	*	*	*	*	*
TX	55	27	18	39	39	23
VA	65	23	12	41	35	24
WA	66	23	11	23	37	39
WI	37	34	29	22	27	51
WV	39	31	30	32	33	36
WY	37	19	44	26	26	48

NOTES: Information in this table was reported by the sites in July and August 2019 and covers Class 50. Some numbers do not sum to 100 percent due to rounding.

* = did not report.

Table 2.7
Percentage of ChalleNGe Graduates in Pre- and Post-TABE Total Battery Grade-Equivalent, by Site (Class 51)

Site	Pre-TABE (%)			Post-TABE (%)		
	Elementary (Grades 1–5)	Middle School (Grades 6–8)	High School (Grades 9–12)	Elementary (Grades 1–5)	Middle School (Grades 6–8)	High School (Grades 9–12)
All sites	52	26	22	21	28	52
AK[a]	N/A	N/A	N/A	N/A	N/A	N/A
AR[a]	N/A	N/A	N/A	N/A	N/A	N/A
CA-DC	*	*	*	*	*	*
CA-LA	78	13	9	12	33	55
CA-SL	49	25	26	15	40	45
D.C.	59	22	20	37	29	34
FL	34	38	28	10	28	62
GA-FG	71	14	15	44	33	24
GA-FS	40	38	23	3	20	77
GA-MV	42	31	26	40	24	36
HI-BP	62	19	19	27	37	37
HI-HI	67	16	17	60	30	10
ID	28	26	46	6	23	71

Table 2.7—Continued

Site	Pre-TABE (%)			Post-TABE (%)		
	Elementary (Grades 1–5)	Middle School (Grades 6–8)	High School (Grades 9–12)	Elementary (Grades 1–5)	Middle School (Grades 6–8)	High School (Grades 9–12)
IL	49	31	20	23	35	42
IN[a]	N/A	N/A	N/A	N/A	N/A	N/A
KY-FK	58	38	5	54	27	20
KY-HN	96	4	0	52	30	19
LA-CB	63	19	18	7	16	76
LA-CM	43	31	26	14	28	58
LA-GL	39	32	29	21	24	55
MD	60	21	18	17	26	57
MI	65	19	16	43	29	28
MS	38	33	29	6	19	75
MT	43	24	33	26	35	40
NC-NL	50	30	20	34	31	35
NC-S	50	31	19	21	40	39
NJ[a]	N/A	N/A	N/A	N/A	N/A	N/A
NM[a]	N/A	N/A	N/A	N/A	N/A	N/A
OK	48	26	27	35	16	49
OR	36	33	31	13	38	49
PR	73	19	7	0	0	100
SC	47	32	22	29	34	36
TN	*	*	*	*	*	*
TX	39	33	27	37	33	30
VA	56	16	27	33	35	33
WA	54	29	17	13	43	44
WI[a]	N/A	N/A	N/A	N/A	N/A	N/A
WV[a]	N/A	N/A	N/A	N/A	N/A	N/A
WY	32	18	50	18	29	53

NOTES: Information in this table was reported by the sites in July and August 2019 and covers Class 51. Some numbers do not sum to 100 percent due to rounding.

* = did not report.

N/A = not available.

[a] These sites (AK, AR, IN, NJ, NM, WI, and WV) used TABE Survey Form 11/12, which does not include a battery test.

Table 2.8
Core Component Completion—Responsible Citizenship, ChalleNGe Graduates (Class 50)

Site	Eligible to Vote	Registered to Vote	Percentage Eligible Who Registered	Eligible for Selective Service	Registered for Selective Service	Percentage Eligible Who Registered
All sites	1,169	1,090	93	1,376	1,351	98
AK	38	37	97	25	24	96
AR	21	20	95	40	39	98
CA-DC	28	28	100	28	28	100
CA-LA	22	22	100	23	23	100
CA-SL	42	31	74	34	24	71
D.C.	9	9	100	9	9	100
FL	40	40	100	29	29	100
GA-FG	51	0	0	81	81	100
GA-FS	46	46	100	35	35	100
GA-MV	32	32	100	28	28	100
HI-BP	94	94	100	72	72	100
HI-HI	62	62	100	42	42	100
ID	13	13	100	32	32	100
IL	19	19	100	19	19	100
IN	5	0	0	29	29	100
KY-FK	10	10	100	10	10	100
KY-HN	14	14	100	15	15	100
LA-CB	32	32	100	94	94	100
LA-CM	21	21	100	57	57	100
LA-GL	33	33	100	26	26	100
MD	24	24	100	71	71	100
MI	22	22	100	32	32	100
MS	39	39	100	54	54	100
MT	17	17	100	29	29	100
NC-NL	44	44	100	18	18	100
NC-S	87	87	100	15	15	100
NJ	11	11	100	6	6	100
NM	20	20	100	52	52	100
OK	13	13	100	36	36	100
OR	37	37	100	57	57	100
PR	51	51	100	42	42	100

Table 2.8—Continued

Site	Eligible to Vote	Registered to Vote	Percentage Eligible Who Registered	Eligible for Selective Service	Registered for Selective Service	Percentage Eligible Who Registered
SC	15	15	100	12	8	67
TN	10	0	0	9	0	0
TX	10	10	100	7	7	100
VA	25	25	100	46	46	100
WA	37	37	100	61	61	100
WI	23	23	100	55	55	100
WV	35	35	100	31	31	100
WY	17	17	100	15	15	100

NOTE: Information in this table was reported by the sites in July and August 2019 and covers Class 50.

Table 2.9
Core Component Completion—Responsible Citizenship, ChalleNGe Graduates (Class 51)

Site	Eligible to Vote	Registered to Vote	Percentage Eligible Who Registered	Eligible for Selective Service	Registered for Selective Service	Percentage Eligible Who Registered
All sites	992	926	93	1,400	1,368	98
AK	30	29	97	23	20	87
AR	20	19	95	52	51	98
CA-DC	23	23	100	23	23	100
CA-LA	30	30	100	34	34	100
CA-SL	32	29	91	25	23	92
D.C.	10	10	100	8	8	100
FL	38	38	100	35	35	100
GA-FG	35	0	0	72	71	99
GA-FS	42	42	100	36	36	100
GA-MV	32	32	100	21	21	100
HI-BP	79	79	100	56	56	100
HI-HI	59	59	100	41	41	100
ID	26	26	100	37	37	100
IL	21	21	100	21	21	100
IN	3	0	0	24	24	100
KY-FK	16	16	100	16	16	100
KY-HN	7	7	100	7	7	100

Table 2.9—Continued

Site	Eligible to Vote	Registered to Vote	Percentage Eligible Who Registered	Eligible for Selective Service	Registered for Selective Service	Percentage Eligible Who Registered
LA-CB	28	28	100	93	93	100
LA-CM	22	22	100	51	50	98
LA-GL	29	25	86	23	19	83
MD	17	17	100	98	98	100
MI	20	20	100	25	25	100
MS	40	40	100	79	79	100
MT	23	23	100	28	28	100
NC-NL	21	21	100	38	38	100
NC-S	24	24	100	19	19	100
NJ	16	16	100	14	14	100
NM	16	16	100	40	40	100
OK	6	6	100	13	13	100
OR	22	22	100	72	72	100
PR	50	50	100	41	41	100
SC	16	8	50	15	6	40
TN	12	2	17	10	0	0
TX	16	16	100	23	23	100
VA	16	15	94	32	32	100
WA	39	39	100	71	71	100
WI	18	18	100	44	44	100
WV	35	35	100	37	36	97
WY	3	3	100	3	3	100

NOTE: Information in this table was reported by the sites in July and August 2019 and covers Class 51.

Community service does vary somewhat across sites; each cadet contributes 40 to 130 hours of community service.

- **Physical fitness (Figures 2.6 and 2.7).** We report one-mile run times and push-ups for Classes 50 and 51. Cadets were able to perform more than 15 additional push-ups and ran about two minutes faster at the end of ChalleNGe. In the 2018 report (Wenger, Constant, and Cottrell, 2018), we included more-detailed data on changes in cadets' BMI and on the proportion of cadets achieving various levels of fitness. In the interest of brevity, we do not include similar information in this report. However, we did collect and analyze the information, and we found results on these health- and fitness-related outcomes that were very similar to previous results.

Figure 2.4
Core Component Completion—Community Service, ChalleNGe Graduates (Class 50)

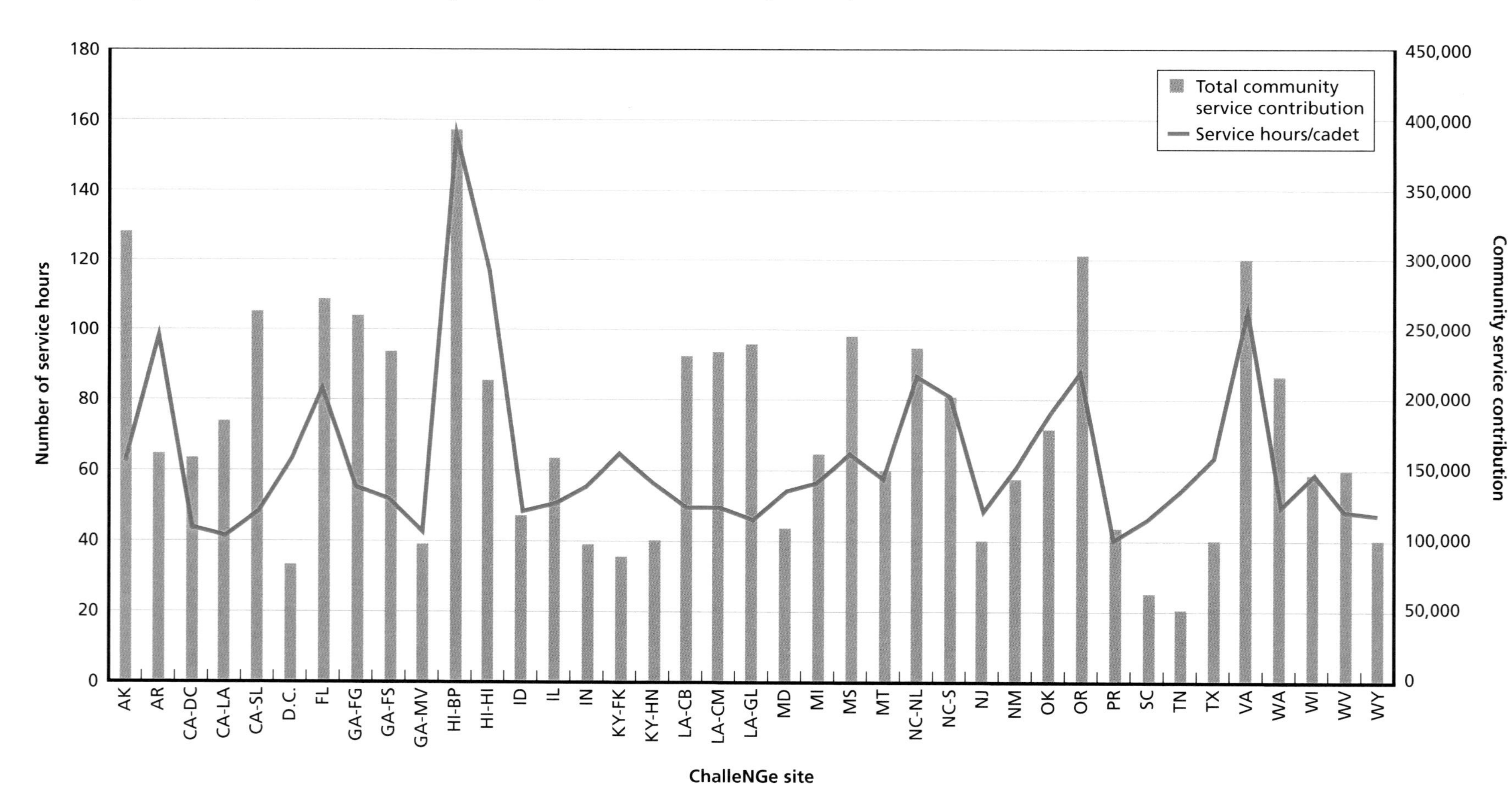

NOTES: Information in this figure was reported by the sites in July and August 2019 and covers Class 50, which began in 2018. The value of community service is calculated using published figures at the state level for 2018 that are available online (Independent Sector, undated). The value of community service was calculated in the same manner in the previous annual reports (Wenger, Constant, and Cottrell, 2018; Wenger et al., 2017; National Guard Youth ChalleNGe, 2015).

Figure 2.5
Core Component Completion—Community Service, ChalleNGe Graduates (Class 51)

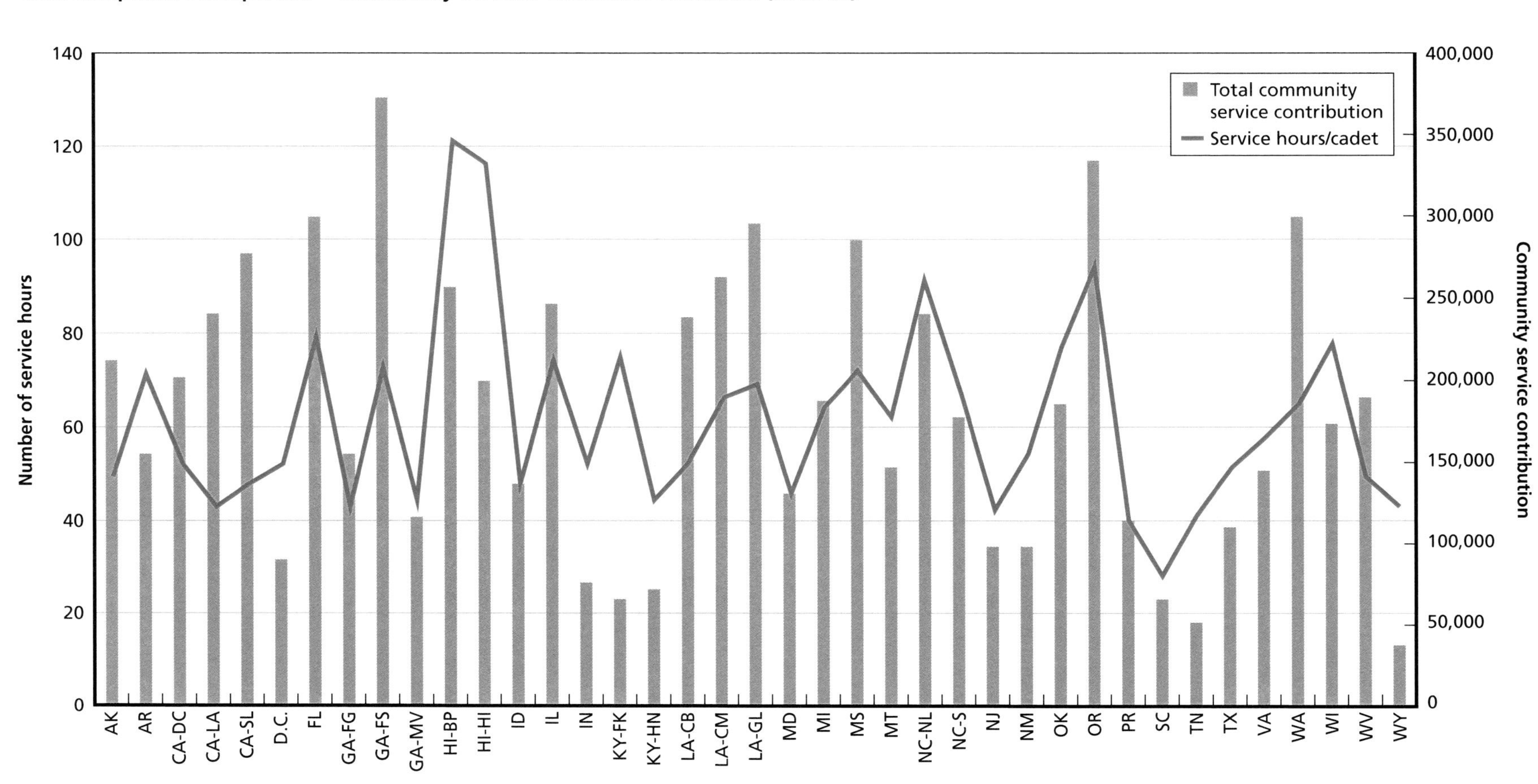

NOTES: Information in this figure was reported by the sites in July and August 2019 and covers Class 51, which began in 2018. The value of community service is calculated using published figures at the state level for 2018 that are available online (Independent Sector, undated). The value of community service was calculated in the same manner in the previous annual reports (Wenger, Constant, and Cottrell, 2018; Wenger et al., 2017; National Guard Youth ChalleNGe, 2015).

Figure 2.6
Residential Performance—Physical Fitness as Measured by the Average Number of Initial and Final Push-ups Completed and Initial and Final Run-Times for Graduates, per Site (Class 50)

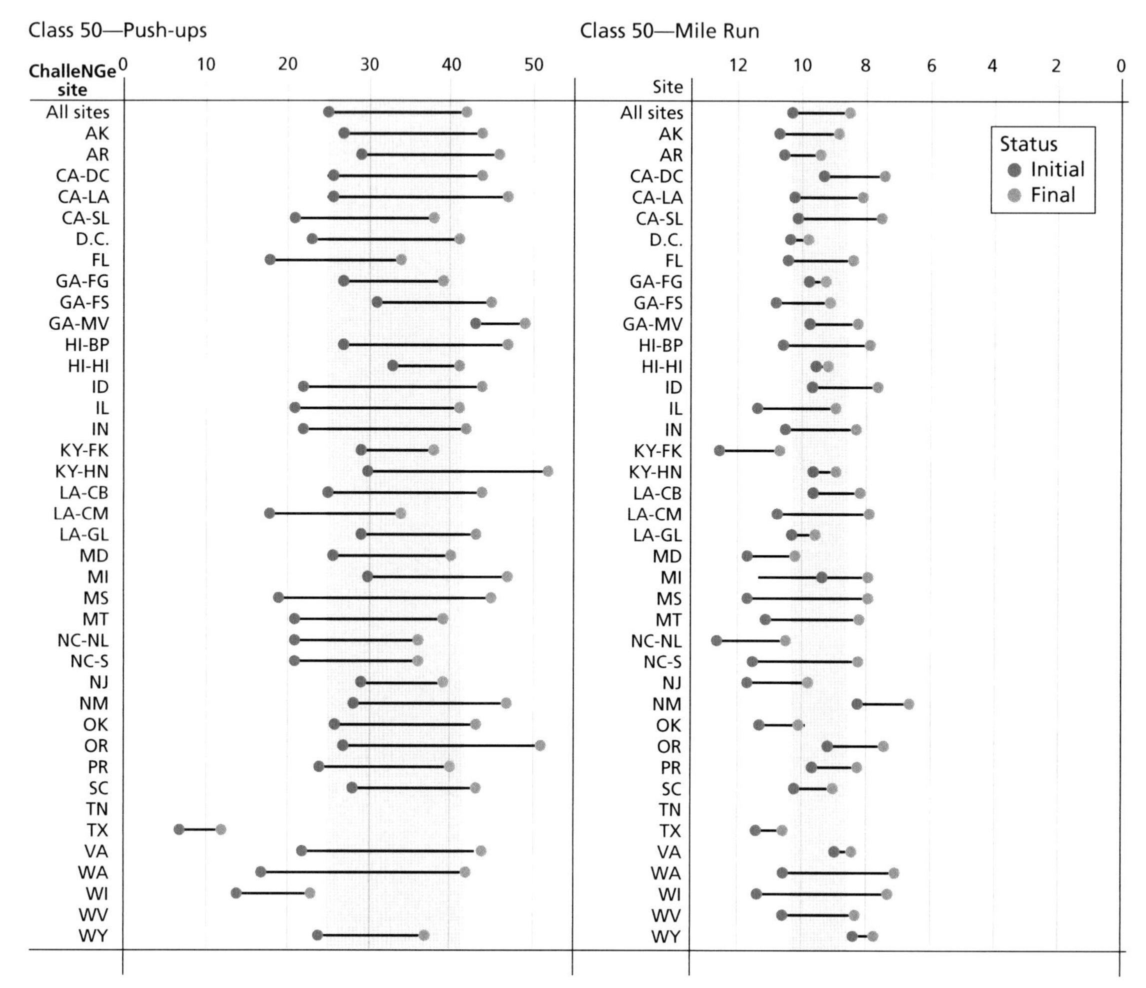

NOTES: Information in this figure was reported by the sites in July and August 2019 and covers Class 50. Texas Class 50 reported pull-ups. Tennessee did not report push-ups and mile run. West Virginia did not report push-ups.

Much of the information in this section has been documented consistently across multiple annual reports. On many metrics (e.g., test score gains and graduation rates), programs appear to be performing much as they did in the past. However, these tables also amply demonstrate the substantial variation that exists across programs. Some of this variation is related to program size, but other metrics, such as test scores and numbers of credentials awarded, are not obviously related to program size. And comparing some of the tables in this chapter with information in past reports suggests that there are trends in the overall number of cadets in several programs and, perhaps, in the graduation rates of cadets and other metrics. At the end of this chapter, we present some information on trends in terms of the total numbers of overall participants and graduates, as well as graduation rates.

In the following section, we present a detailed analysis of a key education-related measure: performance on the TABE. Recent changes to the TABE have implications for ChalleNGe sites; in particular, the gradual adoption of a new version of the test means that comparing

Figure 2.7
Residential Performance—Physical Fitness as Measured by the Average Number of Initial and Final Push-ups Completed and Initial and Final Run-Times for Graduates, per Site (Class 51)

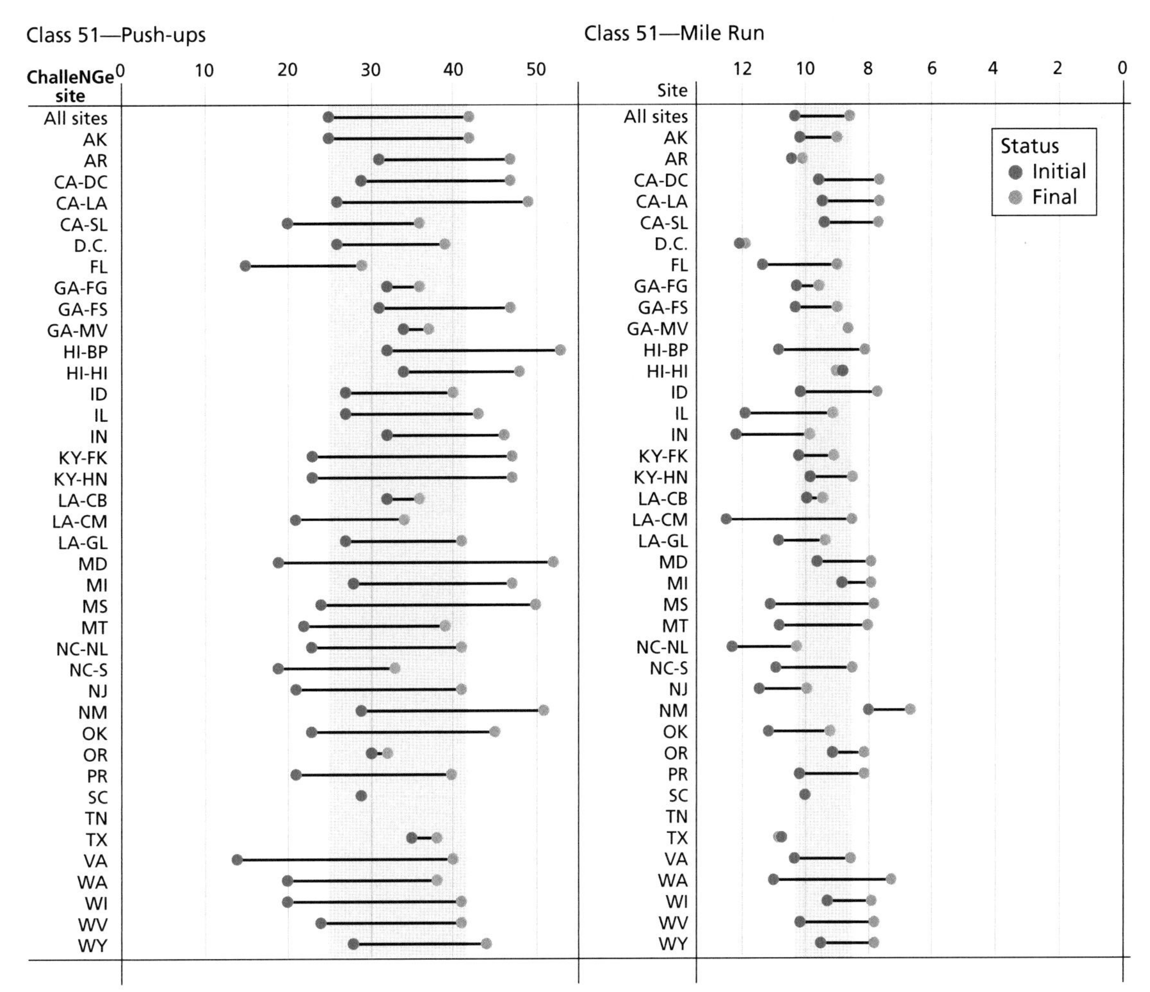

NOTES: Information in this figure was reported by the sites in July and August 2019 and covers Class 51. Tennessee did not report push-ups and mile run. South Carolina did not report final push-ups and final run.

TABE scores across years could be misleading. Our discussion of the TABE and its recent changes is followed by information on program staffing, turnover, and pay, as well as details of cadet placements. In the last section of this chapter, we examine some trends across time.

Tests of Adult Basic Education Scores

The TABE is a standardized test with subtests that focus on reading/language arts and math. TABE is most commonly used in adult basic and secondary education programs.[4] At a mini-

[4] For more information about TABE and the common uses of the test, see U.S. Department of Education, Office of Career, Technical, and Adult Education, Division of Adult Education and Literacy (2016). For more information about ChalleNGe's use of the TABE and the differences between grade equivalents and gain scores, see Wenger et al. (2017) and Wenger, Constant, and Cottrell (2018).

mum, cadets at ChalleNGe take the TABE at the beginning of the program and again at the end of the Residential Phase; some sites also use the TABE more extensively to track progress during the course of the five-plus-month Residential Phase.[5]

The TABE was updated in 2017 to reflect changes in educational standards. TABE developers consider federal legislation, high school testing requirements, as well as established educational standards in the development of their assessments (Data Recognition Corporation [DRC], 2019a). The 2010 release of the K–12 education Common Core State Standards (CCSS) led to identification of new College and Career Readiness (CCR) standards for Adult Basic Education (ABE); this in turn reshaped the instructional focus of many ABE programs and introduced the need for an updated assessment that reflected these new CCR standards. TABE 11/12 was created to align with these standards.[6] Because the new test differs from the previous version (TABE 9/10), this update has implications for ChalleNGe sites. Next, we describe the changes to the TABE in more detail; in the following subsection, we present our analyses of the current TABE data.

Recent Changes to the TABE

TABE 11/12 was adopted by seven ChalleNGe sites during 2018 (Class 51); in the near future, all sites will begin using the TABE 11/12. But during 2018, most sites used the previous version of the TABE, referred to as the TABE 9/10. In the case of the TABE 9/10, scores can be linked to outcomes of interest; for example, a grade-equivalent score of 9.0 is associated with a 70-percent passing rate on the reading, language arts, and math computation sections of the GED, while an 11.0 grade-equivalent score is associated with an 85-percent passing rate on the same tests.[7] Therefore, in past reports (and in this one), we report and analyze TABE information in terms of key grade-equivalent scores. TABE 11/12 does not provide grade equivalents. To ease the transition away from grade equivalents, the TABE developers provide a crosswalk to grade levels (e.g., fifth grade, 11th grade) for TABE 11/12; however, the relationship from scale scores to grade levels has changed between TABE 9/10 and TABE 11/12. Therefore, the relationship between the TABE 11/12 and some outcomes of interest is not yet known.

Completing the move to a new version of the TABE assessment comes with notable changes that will affect the analysis of TABE data that is commonly included in annual ChalleNGe reporting. In Appendix B, we describe exactly how the TABE 11/12 differs from the TABE 9/10. Here, we summarize the changes and focus on what these changes mean for reporting of scores, as well as on critical considerations for comparing data from TABE 9/10 to TABE 11/12.

To continue to use the TABE, ChalleNGe staff will need to understand the changes to the test format and length (see Appendix B). But changes to the test's content area and scoring have the potential to cause confusion in interpretation of the results.

---

[5] According to the data we collected in fall 2018, nine of the 40 ChalleNGe sites reported routinely administering the TABE more than twice during the Residential Phase.

[6] TABE 11/12 was released in September 2017 and is authorized for use through September 2024 (DRC, 2017). The former version of the assessment (i.e., TABE 9/10), authorized in 2003, expired on February 2, 2019. However, a sunset period ending June 30, 2019, was permitted ("Tests Determined to Be Suitable for Use in the National Reporting System for Adult Education," 2018).

[7] For example, see National Reporting Service for Adult Education (2015); Comprehensive Adult Student Assessment System (2016); Comprehensive Adult Student Assessment System (2003); West Virginia Department of Education (undated); Olsen (2009); Wenger, McHugh, and Houck (2006); and Wenger et al. (2017).

The TABE 11/12 scores cover three content areas—reading, language, and mathematics. In contrast to the TABE 9/10, the TABE 11/12 does not provide an overall (battery) composite score. Therefore, we do not report Total Battery scores for the programs that have already shifted to the TABE 11/12 in this report, and we will not include Total Battery scores in future reports.

In contrast to past tests, the TABE 11/12 provides only scale scores, no grade-equivalents. (In the past, programs have used grade-equivalents to track progress.) It is possible to align the new scores with grade levels but not grade equivalents; DRC (the developers of the TABE) has released guidance to align the new scale scores with grade levels, and we use this guidance in our analyses in this report.[8]

However, to date there is no clear guidance on how scale scores on TABE 9/10 correlate with scale scores on TABE 11/12. Based on our analyses, we are confident that scale scores are *not* directly equivalent across the assessments. In other words, a scale score of 500 on TABE 9/10 does *not* represent the same level of performance as a scale score of 500 on TABE 11/12. We determine this by looking at the grade levels to which a 500 scale score equates for each version of the assessment (see Table B.1, in Appendix B, for a crosswalk of reading TABE grade levels on both TABE 9/10 and TABE 11/12). On the TABE 9/10, a reading scale score of 500 is aligned with a fifth-grade level of performance; on TABE 11/12, the same scale score represents performance at a third-grade level. We also know that the possible range of scale score values is not the same across the two assessments, which suggests that the data are not directly comparable.

While there is not yet explicit documentation or research to show that TABE 11/12 is anchored to more challenging academic standards, Pimental (2017) suggests that the CCSS to which the CCR standards are aligned are "ambitious" and that there is likely to be some concern about expecting adult education programs to be even more demanding academically than they were already. Consistent with our findings, this suggests that ChalleNGe programs should be prepared for cadet performance to appear lower on TABE 11/12 than it was on TABE 9/10. Importantly, this change in performance should not be mistaken for an absolute change (i.e., reduction) in cadet skills, abilities, or competencies. Rather, lower levels of cadet performance on TABE 11/12 reflect the change in the assessment against which cadet skills, abilities, or competencies are now being measured. For example, it might be possible that a cadet assessed on TABE 9/10 today would demonstrate performance at an 11th-grade level on that assessment, but, if tested tomorrow using TABE 11/12, the cadet would demonstrate performance at an eighth- or ninth-grade level on the new assessment.[9] The cadet did not necessarily regress academically overnight; the standards against which the cadet is being assessed are more challenging.

As more research becomes available on TABE 11/12 and how performance levels have changed from TABE 9/10 to TABE 11/12, we will provide ChalleNGe programs with the context needed to better understand these changing scores. Moreover, DRC has acknowledged that there is forthcoming research that will link performance on TABE 11/12 to high school equivalency (e.g., GED, TASC) performance (DRC, 2019b). This research will help programs identify which cadets may be academically prepared for a high school equivalency exam and, thus, likely capable of completing a meaningful credential during ChalleNGe or in the Post-Residential Phase.

---

[8] See DRC, undated.

[9] This example is purely for demonstration purposes and does not represent a legitimate crosswalk from TABE 9/10 to TABE 11/12 scores.

TABE Scores, Classes 50 and 51

In this effort, we requested that sites report scores from Math, Reading, and Total Battery.[10] At the beginning of 2018, all ChalleNGe sites used the TABE 9/10, but by Class 51, some sites had moved to TABE 11/12. As noted above, the scores on the two versions of the test are not comparable. Therefore, we present information separately by test version.

Figure 2.8 documents changes in the TABE Reading test over the Residential Phase of the ChalleNGe program, among programs that used the TABE 9/10. These results generally are consistent with findings in our previous reports.[11] At the beginning of ChalleNGe, about one-third of cadets score at the ninth-grade level or higher; by the end of the Residential Phase, more than half of cadets are scoring at or above the ninth-grade level. This suggests substantial progress and indicates that many of these cadets are quite likely to be able to pass the GED test. Figure 2.8 summarizes the information found in Tables 2.4 and 2.5.

Figure 2.9 presents similar information, but for only the small number of sites that had begun using the TABE 11/12 during 2018. In this case, cadets' scores are substantially lower than was the case in Figure 2.1. Less than 15 percent of cadets initially scored at or above

Figure 2.8
Cadet Scores on TABE 9/10 Reading Test Show Substantial Improvement (Between Beginning and End of Residential Phase)

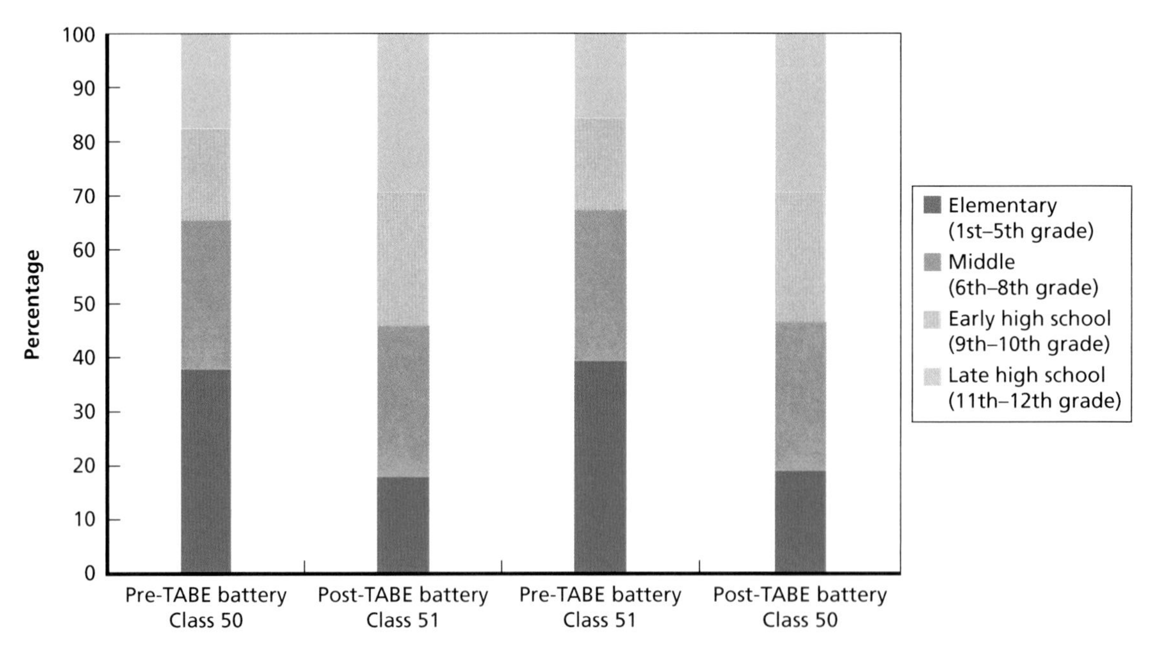

NOTES: This figure is based on information reported by the sites in July and August 2019 and covers only graduates from Classes 50 and 51 who had pre-TABE and post-TABE scores. N = 8,389. Every site except TN reported pre- and post-TABE information.

[10] The Total TABE Battery is formed from the scores on reading, language arts, math computation, and applied math. In an effort to minimize burden on the sites, we collected data on the Total Battery, Reading, and Math Computation tests only; note that these subject tests are the ones that have been found to be predictive of performance on the GED test. In past data collection efforts, we also collected data on the Language Arts subtest, but we found that the Reading subtest provided very similar information. As noted above, the TABE 11/12 does not include Total Battery scores.

[11] We focus on reading here, but the patterns in math and Total Battery scores among sites using TABE 9/10 are similar to those presented in Table 2.1 (and similar to those documented in our previous reports).

the ninth-grade level; by the end of the Residential Phase, less than one-quarter scored at the ninth-grade level. Figure 2.9 still indicates growth over the course of the program, but this figure also suggests that cadets will score lower on the TABE 11/12 than on the TABE 9/10 (this is consistent with information presented in the previous section). Linking the TABE 11/12 to high school equivalency performance will help staff interpret scores, but in the interim, we encourage ChalleNGe staff to use caution in interpreting TABE 11/12 scores.[12] In particular, scoring lower on the TABE 11/12 than on the TABE 9/10 should not be assumed to represent academic regression.

Staffing

Along with information on cadets, the template administered to sites collected information on the number of staff by position and the number of staff by position newly hired in the last 12 months. This measure of staff turnover could indicate dissatisfaction with one or more aspects

Figure 2.9
Cadet Scores on TABE 11/12 Reading Test Also Show Improvement, but Scores on TABE 11/12 Are Lower Than Scores on TABE 9/10

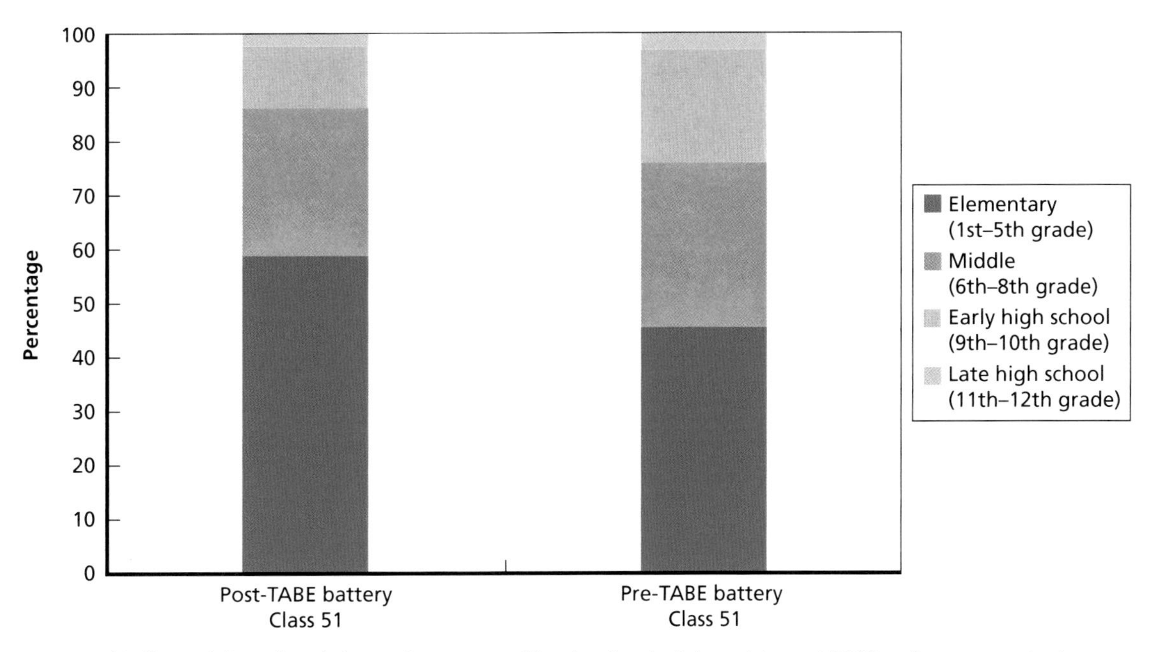

NOTES: This figure is based on information reported by the sites in July and August 2019 and covers graduates from Classes 50 and 51. During Class 51, seven sites had moved to the TABE 11/12. This figure includes data on the Class 51 graduates from those sites with pre- and post-TABE scores reported. N = 733 cadets.

---

[12] If cadets at the sites that have shifted to TABE 11/12 tend to have lower scores than others, this *sample selection* could contribute to the differences in Figures 2.1 and 2.2. However, when we compared 2017 TABE scores of the programs currently using the TABE 11/12 with the scores of all other programs, we found that the sites that currently use TABE 11/12 had average scores at or above other sites (in 2017, all sites used the TABE 9/10). This suggests that moving to the TABE 11/12 will result in lower test scores.

of the job. Compensation is *one* factor that could be related to staff turnover.[13] Below, we describe the staffing at a typical site and provide some information on the relationship between pay and staff turnover.

Staff are categorized into several groups. *Cadre* make up the largest category of ChalleNGe staff members; at all sites, cadre are present around the clock, as they oversee the quasi-military aspect of the program and the cadets. The next-largest group is the *administrators*, which typically consists of the director; deputy director; other members of the leadership team, including the Commandant, who oversees the cadre, data entry staff, budget officer, logistics staff, and other support office staff. *Instructors* deliver the GED or credit recovery curriculum, and they may also administer the TABE. *Case managers* will generally work with the cadets during the Residential Phase to complete their Post-Residential Action Plan (P-RAP), which outlines their goals and the steps they will take to achieve them, and they are also the individuals to whom mentors report placement information on cadets during the Post-Residential Phase. *Recruiters* are responsible for school and community outreach to identify and recruit applicants and generally handle the application process from start to finish. *Counselors*, who are typically licensed, advise on and address the mental health needs of cadets. They also advise and train staff on mental health issues concerning cadets. The *other* category includes medical staff and other program staff who could not be placed into any of the previously listed groups.[14]

Figure 2.10 provides the number of staff, by position, at a typical ChalleNGe program. Nearly half of all staff members are classified as cadre. Given the 24-hour, seven-day-a-week staffing model, this is not surprising. There are some differences in these patterns by program size—of course, larger programs have more staff, but cadre make up a smaller proportion of the staff at the largest programs. This, too, likely reflects the necessity of scheduling cadre around the clock (and conversations with program staff suggest that this is the case). Administrators make up a relatively large share of staff at a typical ChalleNGe program, as do instructors.

Staff turnover is not unusual at ChalleNGe sites. At a typical site, roughly 15 percent of administrators and instructors have been employed for less than 12 months. But among cadre, the turnover rate is twice as high—a typical site reports that about 30 percent of cadre have been employed for less than 12 months.[15] We documented generally similar turnover rates among all sites in our previous report (Constant et al., 2019).

Many factors surely contribute to worker turnover, but ChalleNGe directors and others in leadership positions frequently mention pay as a contributing factor, especially among cadre. In Figure 2.11, we separate programs by cadre turnover. At the programs with the highest levels of cadre turnover, at least half of all cadre had been hired within the 12 months preceding our data collection. At these sites, cadre starting pay was substantially lower than the starting pay

[13] Turnover could occur for other reasons. For example, changes in leadership may result in staff leaving and new staff arriving, or newer programs may have less stability in their staff, as it may take time for the program to develop an identity and select staff with the right fit. Finally, staff turnover can be the result of other natural occurrences, such as retirement or a tendency for short tenure for some positions.

[14] In general, all staff except for cadre work normal working hours and a typical work week. Cadre operate in shifts during the day and on weekends to ensure that there is continuous 24-hour cadet oversight. In a previous data collection effort, we requested information on the part- versus full-time status of staff; most staff are employed full-time.

[15] In calculating these statistics, we excluded sites that had been in operation for less than 12 months.

Figure 2.10
Number of Staff, by Position, Typical ChalleNGe Program

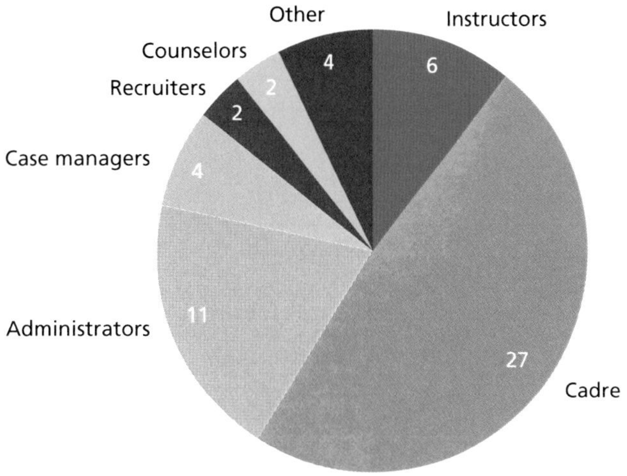

Other 4
Instructors 6
Counselors 2
Recruiters 2
Case managers 4
Administrators 11
Cadre 27

NOTES: This figure is based on information reported by the sites in July and August 2019. The typical number of staff positions is the median number of positions reported among all sites.

Figure 2.11
Starting Pay and Turnover Among Cadre

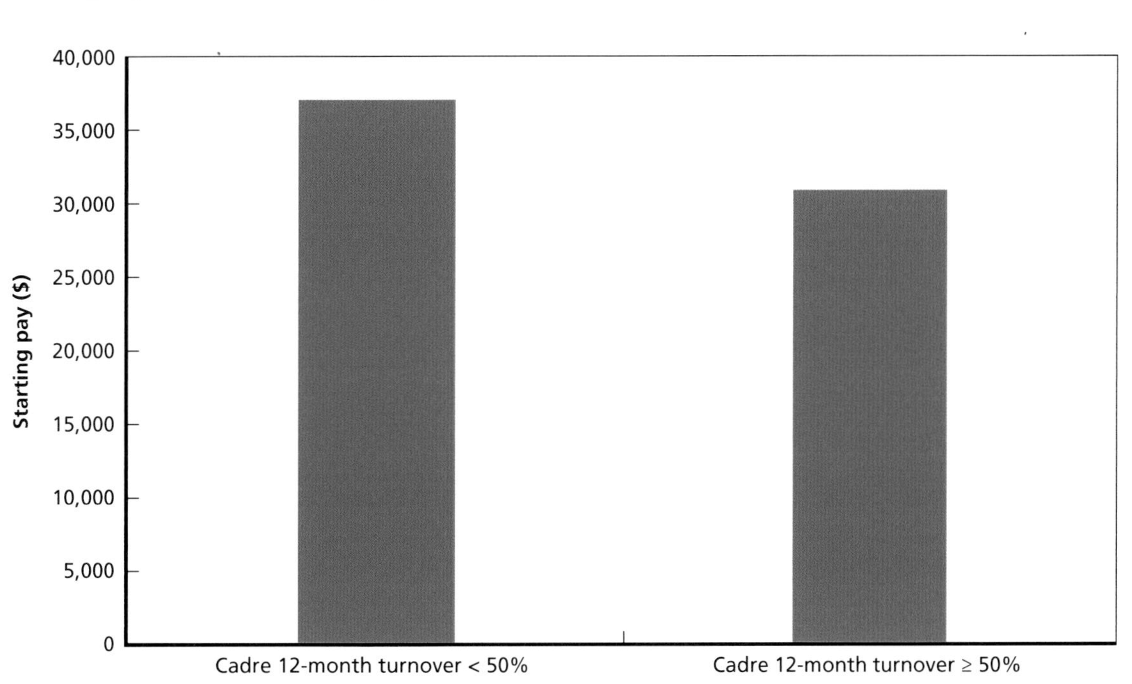

NOTES: This figure is based on information reported by the sites in July and August 2019; we include data on sites established prior to 2017. Cadre turnover exceeded 50 percent at ten sites.

at other sites.[16] Of course, many factors contribute to pay differences, and in many cases sites have relatively little ability to change cadre pay (at least in the short run), but this figure suggests that starting pay is related to turnover among cadre.

When we examined turnover data on instructors, we found similar results (although overall turnover rates are lower among instructors than among cadre). About 40 percent of programs reported instructor turnover of at least 33 percent per year, and, as is shown in Figure 2.12, these programs also reported paying substantially lower starting salaries to instructors than other programs.

Again, there are many factors that are likely to influence staff compensation; examples include the local cost of living and the prevailing wages in the area, as well as state restrictions, overall budget and budget guidance, etc. Sites may control only a few of these factors. But Figures 2.11 and 2.12 do indicate that staff turnover is likely to be related to compensation. Thus, programs that have concerns about turnover should work to influence pay. Our previous report (Constant et al., 2019) documented a small but significant relationship between cadre turnover and graduation rates—cadets at programs with lower levels of cadre turnover were somewhat

Figure 2.12
Starting Pay and Turnover Among Instructors

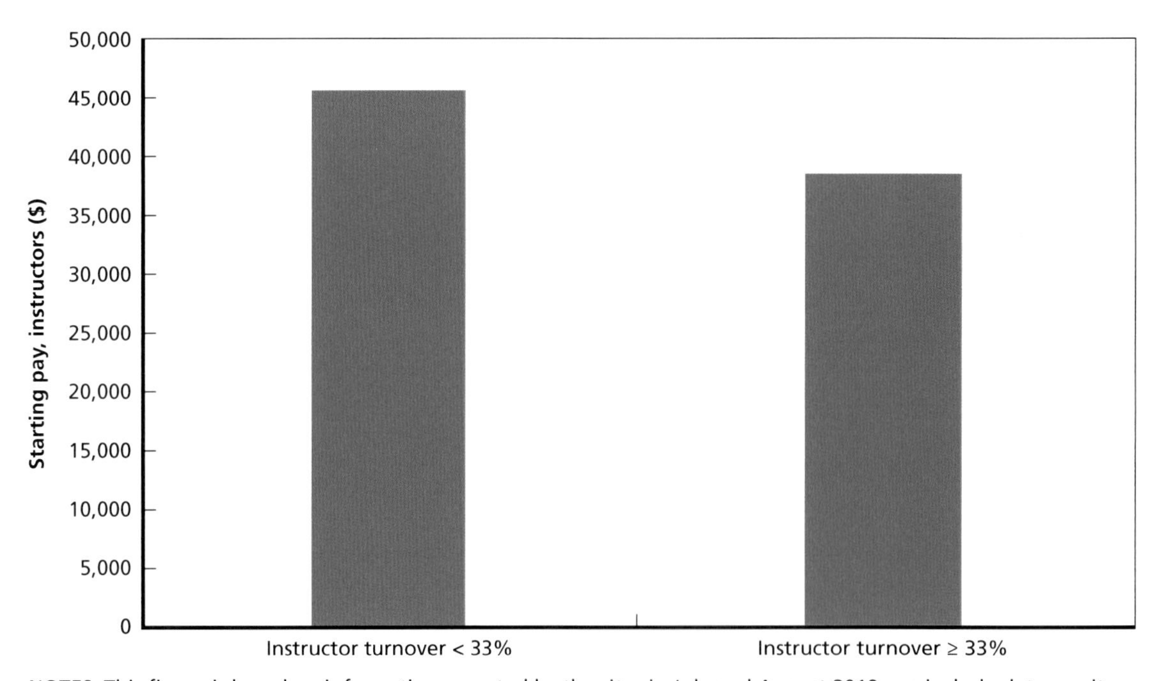

NOTES: This figure is based on information reported by the sites in July and August 2019; we include data on sites established before 2017. Instructor turnover exceeded 33 percent at 14 sites.

[16] The difference in starting pay is roughly $6,000 per year; a *t*-test indicates that the differences are unlikely to occur by chance ($p = 0.02$).

more likely to successfully complete the ChalleNGe program.[17] This provides another motivation to understand how staff respond to pay (and to other working conditions).

Placement

During the course of the ChalleNGe program, all cadets develop a post-ChalleNGe plan. Sites use a specific form, referred to as a P-RAP, as a tool to assist cadets in developing their plans. Plans can be quite detailed and can include additional education, searching for and obtaining a job, joining the military, or some combination of these options. From the perspective of the ChalleNGe program, a *successful placement* is defined as any one or a combination of education, employment, or military participation. As was the case in past data collections, we requested and received information on the placements of recent graduates. At the time of our data collection, graduates of Class 51 (who generally entered ChalleNGe during the latter half of 2018) had, in most cases, completed the program within the prior 12 months. Therefore, we report placement information only at months one and six for these cadets. The placement information is available only for cadets who completed the program and graduated.

Figure 2.13 shows placements of Classes 50 and 51 at three points after graduation. Although nearly one-third of graduates do not have a known placement in the first month after graduation, this figure falls in later months: By months six and 12, nearly 80 percent of gradu-

Figure 2.13
Placements in Months 1, 6, and 12 Among Graduates

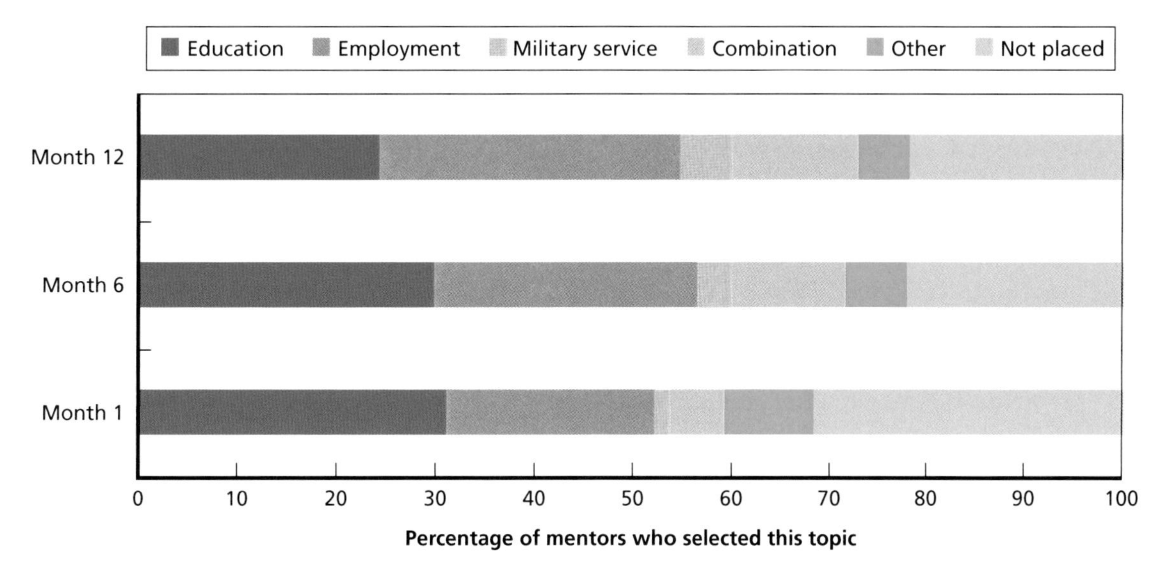

NOTES: This figure is based on information reported by the sites in July and August of 2019 and covers graduates from Classes 50 and 51. Twelve-month placements include only Class 50. "Other" placements include placements that are noted as volunteer or service-to-community positions, as well as placements that are recorded simply as "Other." In each month, there is a substantial number of records with no placement information due to difficulties in contacting cadets or mentors. Rates reported here reflect all information collected from mentors or others and excludes cadets for whom no information is available.

[17] The difference in our previous report occurred between programs with cadre turnover of more than 25 percent versus less than 25 percent. When we use this break point to repeat the analysis shown in Figure 2.4, there is again a difference in salaries, although it is smaller than that shown above.

ates were listed as having a placement. Among cadets who are placed, education and employment are the most common placements. Cadets were most likely to be enrolled in school in the first month after graduation; in later months, cadets were more likely to be employed. The proportion of cadets who reported military service increases over the months, as does the proportion who report some combination of education, employment, and military service. One year after graduation, 5 percent of cadets report serving in the military, and about 13 percent report combining education, employment, and/or military service. At each point, 5 percent to 10 percent of cadets are volunteering or serving their communities in some manner or report a similar activity; these are categorized as *other* (Figure 2.13).

Note that Figure 2.13 does *not* include cadets whose records are missing placement information: Qualitative information gathered during our site visits suggests that some or perhaps even most of these graduates are likely placed, but the programs struggle to obtain placement information on them. One driver of this trend is the difficulty sites encounter when trying to maintain contact with mentors. As noted in Wenger et al. (2017), sites reported spending considerable time and effort trying to maintain contact with mentors. Despite these efforts, at least one-quarter of mentors are no longer reporting to sites by six months after graduation, and reporting rates continue to fall throughout the Post-Residential Phase (Wenger et al., 2017). Among Classes 50 and 51, sites reported information on three-quarters of cadets at the six-month mark and on about two-thirds of cadets at the 12-month mark. Thus, sites are able to obtain placement information on some cadets despite losing contact with the cadets' mentors. According to information gathered during site visits, sites use multiple strategies to contact cadets and document placement. Additional mentor training or other outreach efforts have the potential to improve mentor contact rates; see Chapter Three for information on our ongoing pilot program on interventions to improve mentor reporting rates.

What About Participants Who Do Not Complete ChalleNGe?

As noted in our previous report (Constant et al., 2019) and in the early portion of this chapter, a substantial minority of young people who choose to enter the ChalleNGe program leave the program prior to their graduation date. During our site visits, program staff often discuss the complicated issues that explain these decisions. Program staff have processes in place to help them retain as many participants as possible, but a portion of those who enter Pre-ChalleNGe do not complete the program. Roughly 25 percent of those who enter ChalleNGe do not complete the program successfully. Previous research indicates that females and cadets who are at least 17 years of age upon entry are more likely than others to graduate from ChalleNGe, but the differences are modest in size (see Constant et al., 2019). Earlier research, using multivariate regression models, found similar results in terms of sex and age but also indicated that cadets with higher TABE scores, as well as those from areas with lower levels of poverty, graduated at higher rates—but again, the differences were relatively modest.

In Table 2.10, we present descriptive statistics of all cadets who entered ChalleNGe during 2018; we divide the sample into those who graduated and those who did not, and we present the results by class as well. In *most* cases, the differences between graduates and nongraduates are relatively small. For example, there is no difference in terms of age (even though multivariate regression models and simple descriptive statistics indicate that those who enter at the age of 16 are less likely than older cadets to graduate). However, consistent with earlier research,

Table 2.10
Characteristics of Graduates and Nongraduates, Classes 50 and 51

	Residential Class 50		Residential Class 51	
Characteristic	Graduates	Nongraduates	Graduates	Nongraduates
Age	16.6	16.6	16.5	16.5
Female	24%	20%	24%	22%
White	41%	36%	42%	38%
Black	25%	39%	25%	40%
Latino	20%	14%	20%	12%
Asian	0.6%	0.4%	0.8%	0.3%
American Indian/Alaska Native	3%	3%	3%	3%
Hawaiian or Pacific Islander	2%	1%	2%	1%
Multiracial	7%	5%	6%	5%
Other race/ethnicity	1%	1%	1%	1%
Initial push-ups	24	22	24	25
Initial one-mile run	10:19	10:40	10:19	10:44
Initial BMI	25.2	24.8	25.2	25.4
Individualized education program on file	12%	11%	13%	13%
TABE 9/10				
Pre-TABE reading, elementary	37%	44%	39%	43%
Pre-TABE reading, middle	27%	28%	27%	31%
Pre-TABE reading, early high school	17%	17%	17%	16%
Pre-TABE reading, late high school	18%	11%	16%	11%
TABE 11/12				
Pre-TABE reading, elementary			58%	63%
Pre-TABE reading, middle			27%	29%
Pre-TABE reading, early high school			12%	8%
Pre-TABE reading, late high school			3%	0%

NOTES: We have excluded the following programs from this table because they do not provide information on nongraduates: CA-DC, FL, MD, and SC. We included all available data points on individuals in our database. Class 50 data include 1,704 nongraduates and 4,292 graduates; Class 51 data include 1,518 nongraduates and 4,248 graduates.

female cadets graduate at a higher rate; also, those who enter ChalleNGe with the lowest standardized test scores are less likely than others to graduate, while those who enter with the highest scores are more likely than others to graduate. The most noticeable differences in Table 2.10 are on race/ethnicity. Graduates are more likely than nongraduates to be White/ Caucasian, Hispanic/Latino, or Hawaiian/Pacific Islander. In contrast, nongraduates are more likely to be Black or African-American. Graduation rates among those who identify as Asian, American Indian/Alaska Native, or multiracial are similar to those of cadets on average. These patterns in terms of racial/ethnic differences are interesting and potentially concerning. We would hypothesize that these differences are at least partly due to differences in age, poverty level, educational preparation, and perhaps other site-specific factors. Another possibility is that these effects are particular to 2018 cadets (earlier research found that the differences in ethnicity between ChalleNGe graduates and nongraduates were much smaller; see Wenger et al., 2008). But these differences merit future exploration.

Because a sizable proportion of cadets do not graduate from ChalleNGe, it seems worth considering the likely effects of entering, but not completing, the ChalleNGe program. There is some evidence that suggests that some exposure to ChalleNGe is likely to be beneficial, and some evidence that suggests otherwise. First, the results of a three-year RCT indicate that participants, as a group, gained from taking part in ChalleNGe; most measured gains were related to educational attainment and labor force participation/earnings (see, e.g., Millenky et al., 2011). However, analyses focusing on military performance indicate that those who participated in, but did not graduate from, ChalleNGe were less likely than ChalleNGe graduates to successfully complete their initial term of military service.[18] Additional focus on those who begin, but do not complete, ChalleNGe may suggest interventions that could increase program retention and/or strategies to improve the program admission process.

Time Trends, 2015 to 2018

One focus of this project is to collect consistent, cadet-level data across time. We have begun this process and now have data on eight classes (the classes, two per year, that began in 2015, 2016, 2017, and 2018). Such data are very useful, not only for determining relationships between policies and cadet success, but also for documenting trends over time. Here, we present some time trends in terms of a small number of outcomes—total number of applicants, participants, and graduates. We also track the proportion of cadets who graduate. Due to the recent changes in the TABE and the difficulties with comparing scores from different TABE versions, we do not include any data on TABE scores in this section.

As shown in Figure 2.14, the number of cadets participating in ChalleNGe increased slightly from 2015 to 2017, followed by a slight decline in 2018. Similarly, the total number of

---

[18] See Wenger and Hodari (2004), as well as Wenger et al. (2008). In both cases, first-term retention is higher among military enlistees who completed ChalleNGe than among those who participated in ChalleNGe but did not complete the program. This research focused on ChalleNGe participants who later enlisted in the Army, Navy, Air Force, or Marine Corps.

Figure 2.14
Trends in Applicants, Participants, and Graduates over Time

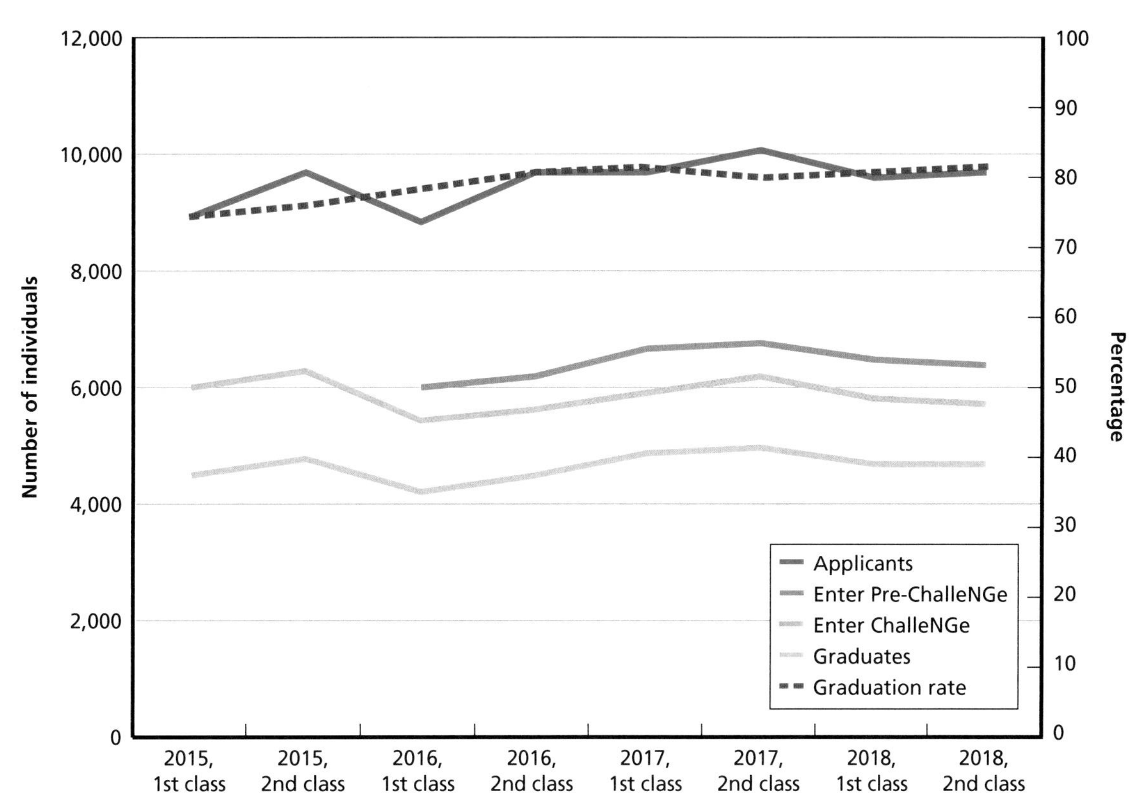

NOTES: Aggregated cadet-level data from each site; this figure is based on information reported by the sites in July and August 2019 for Classes 50 and 51, as well as information collected for previous reports (Constant et al., 2019; Wenger, Constant, and Cottrell, 2018; Wenger et al., 2017). The data included in this figure are not directly comparable to the information included in Figure 2.8 in Wenger, Constant, and Cottrell (2018) because, in our previous report, we excluded data from Puerto Rico due to disruption by Hurricane Maria. This chart includes data obtained later from Puerto Rico. Also, the most recent data in this figure include information on newer sites that were not operational when we began collecting site-level data.

graduates increased from 2015 to 2017, and then declined in 2018. The graduation rate, however, remained roughly constant.[19]

In short, the number of participants and the number of graduates in the ChalleNGe program increased over the past six classes; the metrics on overall graduation rate and TABE scores suggest that the sites were able to increase the number of participants without a marked decrease in graduation rates or achievement, as measured by test scores. During 2018, sites accepted fewer participants and had fewer graduates (although the graduation rate remained roughly constant). This trend seems to be driven by modest decreases in the sizes of some of the larger ChalleNGe programs.

[19] Due to the recent changes in the format of the TABE, we do not include TABE scores in Figure 2.14. However, analyses of the subset of programs that used the TABE 9/10 throughout 2018 indicate that TABE scores among 2018 cadets were similar to those among cadets from earlier years.

Summary

In this chapter, we have documented progress made across the ChalleNGe sites in 2018 and 2019; we focus on the two classes of cadets who began ChalleNGe in 2018. Compared to last year, the implementation of TABE 11/12 represents a substantial change in the standardized testing utilized by the ChalleNGe program. It is not yet clear how to crosswalk scores on this test with other outcomes of interest; the RAND team will provide this information as soon as it is available. Evidence suggests that the TABE 11/12 scores will be lower than those on the TABE 9/10, and staff should not mistake these scores for academic regression. Graduation rates continue to remain much as they have over the past few classes, although enrollment, and therefore the number of graduates, has declined.

An investigation into staff turnover *suggests* a relationship between starting pay and the share of cadre and instructional staff who are new to the program—programs that have fewer new cadre and instructional staff in the past 12 months have higher starting salaries by $6,000 to $7,000. It is important to note that many of our analyses here are descriptive in nature, and it is unclear what drives many of the program differences we report in this chapter. In the next chapter, we describe our other analytic efforts—development of a logic model, progress on site visits, and additional analytic efforts focused on many aspects of the ChalleNGe program. Some of these efforts focus on explaining program-level differences.

CHAPTER THREE

Analyses in Support of ChalleNGe

The ChalleNGe program's documented effectiveness stems from the fact that those who take part in ChalleNGe obtain more education and have better labor market outcomes than similar young people who do not take part in the program; outcomes were measured three years after entering ChalleNGe (Millenky et al., 2011). However, most existing measures of program effectiveness focus on short-term outcomes, such as the placement rate among recent graduates. As was explained in the previous chapter, in the ChalleNGe program, *placement* is defined as participation in the labor market, military enlistment, working toward an education credential, or some combination of these. Existing measures of placement do not record the level of education obtained or the wage rate or other aspects of a graduate's job. Thus, the existing measures lack specificity that would be necessary to determine, for example, the expected earnings of ChalleNGe graduates. Moreover, the measures are short term in nature.

In this chapter, we discuss the logic model, which incorporates the longer-term outcomes of the program. In a previous report, we first introduced a TOC and the logic model (Wenger et al., 2017). We include the logic model in this report and provide a brief discussion of its use to date in the project. We also discuss the implications for the long-term metrics under development. We then describe the status of the site visits, which inform multiple aspects of the overall study objective. In particular, the site visits add context to our interpretation of the data we collect annually from the program sites to support the ChalleNGe program's yearly report to Congress. We also describe a variety of analytic research efforts and pilot projects to support the development of these longer-term metrics.

Following our discussion of the logic model and the status of the site visits, we synthesize the findings from one analytic effort and one pilot project, respectively:

- identifying promising practices in CTE and examining participation in CTE across ChalleNGe sites, as well as the opportunities and constraints to providing it
- designing and incorporating training modules and exercises into an existing ChalleNGe site's mentor training program, drawing on best-practices in mentoring.

We also briefly describe two other ongoing analytic efforts and one additional pilot to support the development of long-term metrics. The two other analytic efforts are examining high-paying job skills and the approaches that ChalleNGe sites are taking to meet the mental health needs of their cadets. The other pilot involves the implementation of an alumni survey to collect information on ChalleNGe graduates from one site. Collectively, these analytic efforts and pilots draw on multiple data sources and approaches—including information collected annually from program sites, qualitative data collected from site visits, and other extant data sources, such as publicly available data from nationally representative surveys.

The analytic efforts and pilots were developed in response to key issues that we identified during site visits and in our direct interactions with program site leadership and staff, as well as with input from our sponsor. They draw on multiple data sources—including information collected annually from the program sites, qualitative data collected from site visits, and other extant data sources, such as publicly available data from nationally representative surveys. The RAND research team identified key issues during the interviews with site personnel. The underlying research questions were developed based on an assessment of the salience of the issues revealed in the interviews, team member expertise, and budget and timeline considerations. The mentoring pilot, for example, was designed based on an analytic effort involving a review of best-practices in mentoring reported on previously. It is important to note that the analytic efforts the RAND team chose to work on by no means represent *all* of the issues facing program sites.

Logic Model

As noted above, we developed a program logic model—which we introduced in the first annual report (Wenger et al., 2017)—to assist in our development of long-term metrics. The model delineates the inputs, processes or activities, expected outputs, and desired outcomes of the ChalleNGe program (for more detail on developing logic models, see Shakman and Rodriguez, 2015).

Program logic models are a useful way of specifying the reasoning behind program structure and activities, as well as how those activities are connected to expected program results (Knowlton and Phillips, 2009). They are used to illustrate how program resources, activities, services (inputs), and direct products of services (outputs) are designed to produce short-term, medium-term, and long-term outcomes. Logic models also identify broader community effects that should result from program activities and services (Knowlton and Phillips, 2009). In this way, these models can communicate how a program contributes not only to the specific needs and outcomes of program participants but also to the broader community and society at large. Program logic models also serve as a blueprint for evaluating how effectively a program is meeting its expected goals.

The ChalleNGe logic model emphasizes the temporal aspects of the ChalleNGe program and its influence on participants, and it lays out expected results in detail (see Figure 3.1). The initial logic model was informed by a review of program documentation and annual reports and by site visits to two ChalleNGe locations (the Mountaineer ChalleNGe Academy in West Virginia and the Gillis Long ChalleNGe site in Louisiana). We have used the ChalleNGe logic model to clarify our thinking about the program's inputs, outputs, and outcomes. We also have used the model to communicate key aspects of ChalleNGe to a variety of stakeholders, including policymakers, program directors, program staff, and other researchers. In each case, the model has been a helpful communication tool. We have made small adjustments to the model based on input received throughout the project and will continue to refine and expand on the current program logic model and its uses in future reports.

At present, the ChalleNGe program sites focus primarily on collecting and reporting on metrics associated with the inputs, activities, and outputs. Metrics on short-term outcomes

Figure 3.1
Logic Model Describing the National Guard Youth ChalleNGe Program

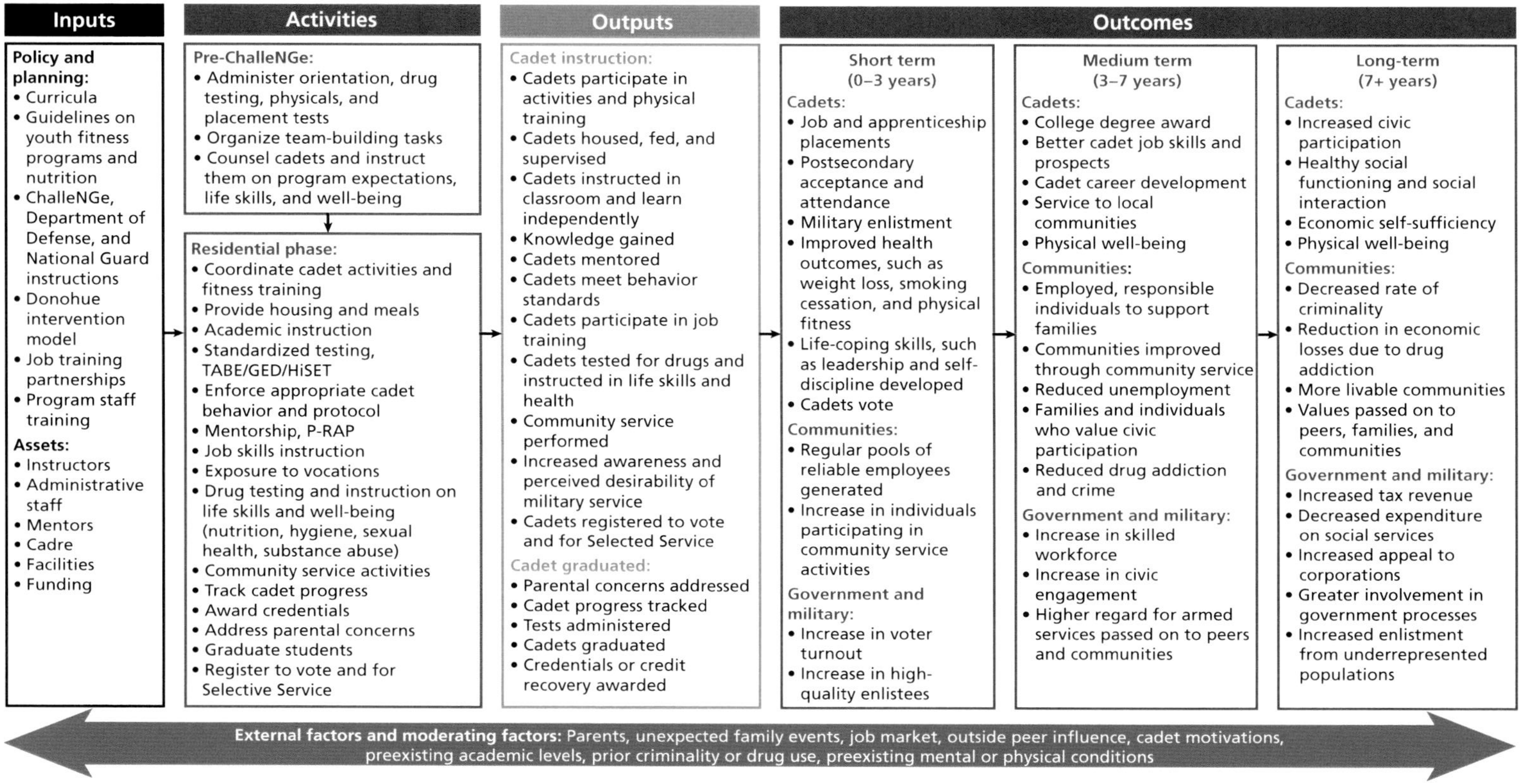

Inputs

Policy and planning:
- Curricula
- Guidelines on youth fitness programs and nutrition
- ChalleNGe, Department of Defense, and National Guard instructions
- Donohue intervention model
- Job training partnerships
- Program staff training

Assets:
- Instructors
- Administrative staff
- Mentors
- Cadre
- Facilities
- Funding

→

Activities

Pre-ChalleNGe:
- Administer orientation, drug testing, physicals, and placement tests
- Organize team-building tasks
- Counsel cadets and instruct them on program expectations, life skills, and well-being

↓

Residential phase:
- Coordinate cadet activities and fitness training
- Provide housing and meals
- Academic instruction
- Standardized testing, TABE/GED/HiSET
- Enforce appropriate cadet behavior and protocol
- Mentorship, P-RAP
- Job skills instruction
- Exposure to vocations
- Drug testing and instruction on life skills and well-being (nutrition, hygiene, sexual health, substance abuse)
- Community service activities
- Track cadet progress
- Award credentials
- Address parental concerns
- Graduate students
- Register to vote and for Selective Service

→

Outputs

Cadet instruction:
- Cadets participate in activities and physical training
- Cadets housed, fed, and supervised
- Cadets instructed in classroom and learn independently
- Knowledge gained
- Cadets mentored
- Cadets meet behavior standards
- Cadets participate in job training
- Cadets tested for drugs and instructed in life skills and health
- Community service performed
- Increased awareness and perceived desirability of military service
- Cadets registered to vote and for Selected Service

Cadet graduated:
- Parental concerns addressed
- Cadet progress tracked
- Tests administered
- Cadets graduated
- Credentials or credit recovery awarded

→

Outcomes

Short term (0–3 years)

Cadets:
- Job and apprenticeship placements
- Postsecondary acceptance and attendance
- Military enlistment
- Improved health outcomes, such as weight loss, smoking cessation, and physical fitness
- Life-coping skills, such as leadership and self-discipline developed
- Cadets vote

Communities:
- Regular pools of reliable employees generated
- Increase in individuals participating in community service activities

Government and military:
- Increase in voter turnout
- Increase in high-quality enlistees

→

Medium term (3–7 years)

Cadets:
- College degree award
- Better cadet job skills and prospects
- Cadet career development
- Service to local communities
- Physical well-being

Communities:
- Employed, responsible individuals to support families
- Communities improved through community service
- Reduced unemployment
- Families and individuals who value civic participation
- Reduced drug addiction and crime

Government and military:
- Increase in skilled workforce
- Increase in civic engagement
- Higher regard for armed services passed on to peers and communities

→

Long-term (7+ years)

Cadets:
- Increased civic participation
- Healthy social functioning and social interaction
- Economic self-sufficiency
- Physical well-being

Communities:
- Decreased rate of criminality
- Reduction in economic losses due to drug addiction
- More livable communities
- Values passed on to peers, families, and communities

Government and military:
- Increased tax revenue
- Decreased expenditure on social services
- Increased appeal to corporations
- Greater involvement in government processes
- Increased enlistment from underrepresented populations

External factors and moderating factors: Parents, unexpected family events, job market, outside peer influence, cadet motivations, preexisting academic levels, prior criminality or drug use, preexisting mental or physical conditions

NOTES: The Donohue intervention model was the initial design and description of the ChalleNGe program (Price, 2010). GED and HiSET credentials are awarded based on performance on standardized tests. The P-RAP is designed to support planning and goal development among cadets.

(achieved within one year, during the Post-Residential Phase) are collected and reported on, but the extent and consistency of reporting varies across program sites. Some of the outcome measures included in Figure 3.1 (especially those listed as community outcomes) may be initially influenced by the program participants; for example, ChalleNGe graduates may vote and perform community service at higher levels than would be expected without program participation. However, it is also possible that the program will eventually have a broader influence on community-level outcomes.

Currently, program sites collect information on graduates, not on those who participated in but did not complete the program (nongraduates). Although collecting information on all program participants (graduates and nongraduates) would be preferred, program sites currently have limited ability to do so because they rely on mentors to report information on graduates. Nongraduates are not formally assigned a mentor or tracked. Because sites collect contact information on all participants, sites could consider administering a survey to capture a sample of graduates and nongraduates. The efficacy of a survey to collect placement information is currently being explored as a pilot with one of the program sites (we provide an explanation of the pilot later in this chapter). One concern is that response rates are likely to be low among nongraduates without significant investment of survey administration resources. The treatment group in the MDRC study referenced in Chapter One included those admitted in its "intent-to-treat" design, which meant collecting information on both graduates and nongraduates (Bloom, Gardenhire-Crooks, and Mandsager, 2009; Millenky, Bloom, and Dillon, 2010; Millenky et al., 2011). However, this was a one-time study requiring significant investment of data collection resources that is difficult to replicate on an ongoing basis. Nonetheless, avenues to collect information on both graduates and nongraduates should continue to be explored to provide sites with a more complete picture of their performance.

The logic model has several implications for the ChalleNGe program. As we noted above, ChalleNGe's mission is to produce graduates who are successful, productive citizens in the years after they complete the program. The research that established the effectiveness of ChalleNGe on job performance and earnings, and the cost-benefit calculations associated with that research, focused on longer-term outcomes, which we can refer to as *impacts* (see Chapter One of this report, as well as Wenger et al., 2017). Currently, program sites focus on collecting outputs, which are considered measures of performance, not effectiveness. To measure effectiveness, program sites will need to focus more on collecting outcomes in order to determine the extent to which ChalleNGe is meeting its stated *mission*. Much of the information presented in Chapter Two focuses on inputs and outputs (the left-hand side of the logic model) and, to a lesser extent, on short-term outcomes (the right-hand side of the logic model). Ultimately, measuring the extent to which the program is meeting its mission will require collecting longer-term outcomes, similar to the type collected as part of the RCT described in Chapter One of this report (Bloom, Gardenhire-Crooks, and Mandsager, 2009; Millenky, Bloom, and Dillon, 2010; Millenky et al., 2011). Because of the expense and significant burden on participants, we do not intend to replicate an RCT. As we work to develop longer-term metrics to measure impact, we will also explore methods of estimating the overall returns to the program that do not entail enrolling thousands of ChalleNGe cadets into RCTs. We discuss some of these efforts below.

Site Visits

Over the course of this project, the RAND team visited each of the 39 current ChalleNGe sites. We developed a detailed protocol that includes questions for the director, deputy director, commandant (head of the cadre), recruiter, placement coordinator, mentoring coordinator, lead instructor (or principal), and the management information specialist. The protocol includes questions about the site's core mission, resources, staffing and hiring, outreach efforts, and relationship with the community, as well as many questions about the day-to-day operations and the types of data that are collected on cadets both during and after the program. The protocol also includes questions about recruiting and training mentors, cadet placements, the process of recruiting cadets and mentors, instruction, credentials, and occupational training opportunities. The protocol includes additional questions for sites in states with multiple ChalleNGe programs and for sites in states that offered Job ChalleNGe.[1] Table 3.1 shows the timing of our visits.

To date, we have visited all 39 sites.[2] These site visits have served multiple purposes. During the first months of the project, we visited several sites to gather information as we developed our logic model. Site visits in the interim have been helpful as we worked to refine our research efforts in support of developing long-term metrics of program effectiveness. From our site visits, we have learned quite a bit about the sources of variation across ChalleNGe sites, which has been helpful in designing our annual data collection instruments. We have also learned about the state and local policy context and the implications of accessing data for mea-

Table 3.1
Schedule of Site Visits

ChalleNGe Site	2016–2017	2018	2019
Alaska		✓	
Arkansas			✓
California-Discovery	✓		
California-Grizzly	✓		
California-Sunburst		✓	
District of Columbia			✓
Florida	✓		
Georgia–Fort Gordon		✓	
Georgia–Fort Stewart		✓	
Georgia-Milledgeville		✓	
Hawaii–Barbers Point	✓		

[1] Job ChalleNGe is a program with a focus on job training that is designed to follow ChalleNGe. At the time of our site visits, it was available as a pilot program at only three of the states with ChalleNGe sites: Michigan, South Carolina, and Georgia. Since then, Job ChalleNGe has expanded to several additional states.

[2] We visited the Louisiana–Gillis Long site and the West Virginia site in late 2016 when we were designing our logic model. During the next year, we developed and considerably expanded our protocol; therefore, we returned to both West Virginia and Louisiana–Gillis Long in 2018.

Table 3.1—Continued

ChalleNGe Site	2016–2017	2018	2019
Hawaii-Hilo	✓		
Idaho			✓
Illinois			✓
Indiana	✓		
Kentucky–Fort Knox (Bluegrass)		✓	
Kentucky-Harlan (Appalachian)	✓		
Louisiana–Camp Beauregard		✓	
Louisiana–Camp Minden		✓	
Louisiana–Gillis Long	✓	✓	
Maryland			✓
Michigan		✓	
Mississippi			✓
Montana	✓		
North Carolina–New London		✓	
North Carolina–Salemburg		✓	
New Jersey	✓		
New Mexico	✓		
Oklahoma			✓
Oregon	✓		
Puerto Rico		✓	
South Carolina		✓	
Tennessee			✓
Texas-East			✓
Texas-West			✓
Virginia	✓		
Washington	✓		
Wisconsin	✓		
West Virginia	✓	✓	
Wyoming	✓		

NOTE: The Texas ChalleNGe programs merged in late 2018; we will visit the single Texas program in 2019.

suring long-term metrics. We gathered information on the data maintained at each site and have been able to determine which sites are good candidates for future short-term pilot projects that would enable us to explore aspects of the ChalleNGe model that pose particular challenges, or pilot projects that could inform policies and practices across all sites. For example, at all site visits, program staff have reported varying degrees of difficulty when trying to gather accurate and timely information about graduate placement in the Post-Residential Phase. Sites typically rely on mentors to report this information, which is based on their regular interaction with graduates as their mentees. We are currently working with a site to develop other options for collecting placement data on graduates, such as through social media and alumni events.

We are also working with another site to improve mentor retention and engagement with mentees in the Post-Residential Phase. This effort, described in the next section, is in response to a commonly expressed frustration across sites at the drop-off in mentor-mentee engagement in the three- to six-month period after graduation. We are currently working with the site to identify the mechanisms underlying the drop-off in engagement and ways to address this issue through mentor training and supports.

Description of Analytic Efforts

Along with the annual reports to Congress, the development of the TOC and logic models, and the site visits, we are undertaking a series of analytic research efforts. These analytic efforts were listed in Table 1.1 in Chapter One. Although these efforts jointly support the development of longer-term metrics and the goal of measuring ChalleNGe's effectiveness, each of these efforts is also focused on a specific topic or issue identified through site visits and engagement with program leadership and staff. In this section, we briefly describe three of the analytic efforts and the two pilot projects. As noted above, we use the phrase *analytic effort* to describe research efforts involving information from multiple ChalleNGe sites; *pilot project* refers to efforts developed in close consultation with a single ChalleNGe site. We delve into greater detail regarding the analytic effort on examining CTE practices across sites and with one of the pilots on improving the mentor training program. We also briefly describe the remaining two ongoing analytic efforts and one other pilot.

Examining the Implementation of Career and Technical Education

In last year's report, we described an ongoing study examining CTE in ChalleNGe sites (Constant et al., 2019). CTE has been receiving attention nationally and is an area of interest among ChalleNGe sites, particularly with the advent and expansion of Job ChalleNGe.[3] To respond to this interest, the RAND team sought to identify best-practices in CTE and understand both the opportunities and constraints to providing CTE at ChalleNGe sites. Here, we provide some of the highlights of this study, which will be published separately with its complete findings.

[3] Job ChalleNGe is a 5.5-month residential occupationally focused training program for eligible ChalleNGe graduates. Participants in Job ChalleNGe work toward a job-ready, stackable credential in an occupational field. Job ChalleNGe began with three pilot states in 2016 and has since expanded to include other states with ChalleNGe programs. Job ChalleNGe participants are typically housed on a ChalleNGe campus or close by and take classes at a local community college or other training provider.

To carry out this study, the RAND team conducted the following activities: (1) examined the current literature on CTE to identify promising practices; (2) described CTE practices across ChalleNGe sites by drawing on the data reported in the annual data call and benchmarking those practices with data reported by the National Center for Education Statistics from the High School Longitudinal Study of 2009 (HSLS:09) on public high school student participation in CTE; and (3) conducted interviews with five ChalleNGe sites to understand their perspectives on both opportunities and constraints to providing CTE. Our criteria for selecting the sites to interview was based on examining the data collected from the annual data call to ensure that we spoke with at least one site with a more developed CTE program in terms of the variety of CTE classes offered and the rate of participation of cadets, but also to ensure that we spoke with sites that broadly represented ChalleNGe in terms of program size, geography/location, and years of operation.

Promising Practices in CTE

A literature review identified five promising practices that have been found to be effective in the provision of CTE. These are described in more detail below.

Structured pathways. *Career pathways* are structured sequences of courses and training that are meant to place students on a clearly defined "pathway" toward an occupational credential—be it a certificate, license, associate's degree, or some combination thereof. Several studies point to the benefit to students of engaging in CTE through a clearly defined sequencing of classes, as opposed to dabbling without a clear aim or direction (Castellano et al., 2014; Kemple, 2008; Karp et al., 2007).

Career preparation supports. The combination of formal career pathway coursework with corresponding career preparation activities helps students focus and plan for their careers, especially if they intend to seek employment rather than pursue a bachelor's degree. Several studies have found positive effects when formal career pathways are combined with career preparation activities (Castellano et al., 2014; Karp et al., 2007; Hemelt, Lenard, and Paeplow, 2019).

Work-based learning (WBL). WBL provides structured on-the-job training in the form of an internship or apprenticeship-style experience in which students work under the direct supervision of the participating employer while receiving guidance that integrates and reinforces what is being taught in CTE. WBL should be directly linked and complementary to the CTE curriculum the student is participating in, and research points to the importance of WBL in a CTE program (Bozick and MacAllum, 2002; Bozick et al., 2019).

Integrated academic-occupational curriculum. The explicit linking of concepts and skills taught in academic courses with those taught in occupationally based courses helps students understand how their classes apply to the challenges they may face in the workplace. An integrated academic-occupational curriculum is intended to promote the preparation of students for college even while they are enrolled in a CTE-focused course schedule (Visher and Stern, 2015).

Industry engagement. Although obtaining it is challenging in practice, there is an increasing emphasis on the importance of school and industry engagement to identify careers and skills in demand, keep curricula up to date, and secure WBL opportunities for CTE students (U.S. Department of Labor et al., 2014).

CTE Practices Across Sites

We draw on CTE participation data collected for Classes 48, 49, 50, and 51 and present them in Table 3.2. The data reveal significant variation in CTE participation of program graduates across sites. Overall, 15 percent of graduating cadets in classes 48–51 received CTE credits. In intensive programs, or the sites with at least 50 percent of cadets receiving CTE credits, around 90 percent of cadets received some CTE credits on average, compared with 65 percent in programs with at least some reported CTE offerings. This compares to 88 percent of traditional high school students receiving at least some CTE credits as reported in the nationally representative HSLS:09 data. Given the wide disparity in reported CTE provision between ChalleNGe and the national average, direct comparisons between public high schools and ChalleNGe are more appropriately made with the intensive CTE ChalleNGe sites.

In terms of some of the types of courses offered by those ChalleNGe sites that reported providing CTE, the most commonly provided courses were in beginning construction and related trades (15 ServSafe® food handling sites); courses in industrial manufacturing, welding, and heavy equipment operation (ten sites); auto mechanics (ten sites); Occupational Safety and Health Administration certification (seven sites); ServSafe® food handling and other hospitality services (nine sites); patient care, cardiopulmonary resuscitation, and other health-support services fields (eight sites); and auto mechanics (five sites). Among public high school students, a significant share take CTE courses in family and consumer sciences education (37 percent), communication and communication technologies (29 percent), office support (25 percent), and computer and information sciences (20 percent). Although other courses are less common, significant numbers of public high school students take occupationally focused courses in such areas as hospitality services (13 percent), engineering and design (13 percent), and manufacturing and technology (11 percent) (U.S. Department of Education, Institute of Education Sciences, National Center for Education Statistics, 2019a, and U.S. Department of Education, Institute of Education Sciences, National Center for Education Statistics, 2019b).

Table 3.2
Participation Rates in the HSLS and ChalleNGe (Classes 48–51)

	HSLS	ChalleNGe		
	National Average	All	In Sites with Some CTE Offerings	In Sites with Intensive CTE Credit Programs[a]
Participation (%)				
With any CTE	88	15	65	90
With at least one credit	81	14	59	82
With at least two credits	58	10	43	57
With at least three credits	39	7	29	37
With a CTE credential	—	18	46	64

SOURCES: 2017 and 2018 ChalleNGe data calls and National Center for Education Statistics summary tables.

NOTES: [a] Sites with at least half of their cadets receiving CTE credits. The intensive threshold was defined annually, and sites are included only in the years during which they met the standard. Sites classified as intensive at any point are AK, CA-Grizzly, IN, MI, OK, SC, WA, and WV. MI and OK met the standard only in Classes 48 and 49, while AK, IN, and SC met the standard only in Classes 50 and 51.

ChalleNGe sites are focused on ensuring that cadets are working toward a high school or equivalent credential, such as a GED/HiSET, credit recovery, or high school diploma. CTE was typically approached as a supplement to those academic courses. Each site had developed policies regarding eligibility to participate in CTE, with some CTE courses available to all cadets and others only to cadets meeting certain academic requirements. The selection of CTE courses offered by sites was contingent on several factors: the ability to fit an entry-level basic introduction to the course within the compressed schedule of cadets; availability of course material and instruction either through dual enrollment at a local community college or depending on the availability of a qualified instructor on staff to teach the course; and some consideration of local, regional, and state-level industry and demand for skills. Some sites subscribed to online programs that provided basic, introductory exposure to CTE fields, but without the hands-on practical experience.

Our closer investigation of CTE in the handful of sites we interviewed revealed variation in the implementation of promising practices in CTE across sites. It is important to recognize that sites faced significant challenges in providing CTE during the five-month Residential Phase while cadets were also attempting to complete their academic and other requirements. Moreover, there were logistical and cost constraints associated with transporting cadets to and from their CTE courses, typically in the form of dual enrollment at nearby community colleges. A synthesis of our findings is discussed below.

Several Programs Offered Dual Enrollment to Structure CTE Coursetaking

Three of the five programs that we interviewed offered structured pathways through a dual enrollment program with a local community college. Cadets were able to take courses that could count toward further community college pursuits or a credential if they continued in the same occupational area.

All Programs Offered Career Preparation, but Some Program Sites Supplemented it with Additional Career Placement Supports

Three of the five programs we spoke with in our follow-on interviews supplemented the P-RAP, citing specific activities such as career fairs and mock interviews with members of the local community and businesses. In one program, cadets enroll in a one-credit course designed to assist with career planning and preparation.

Formal Work-Based Learning Was Not Common Among the ChalleNGe Sites We Interviewed

With the exception of one program, formal WBL was not available to cadets. In that one program, approximately 25 percent of cadets participated in formal internships that lasted four hours a day, four days per week, for four weeks, ranging from internships in construction and the automotive industry to retail and restaurant service to administrative positions in offices. These WBL opportunities stem from the program's long-term relationship with the local business community.

An Integrated Academic-Occupational Program Was Not Common Across the ChalleNGe Sites Interviewed

Formal integration of academic and occupational curricula, content, and instruction was not common across ChalleNGe sites. Most of the ChalleNGe sites interviewed rely on community colleges to provide CTE instruction designed and implemented at the college. This makes it difficult for community college instructors and ChalleNGe instructors to work together to integrate the curriculum. Moreover, ChalleNGe sites are using a blend of curricula to meet their GED, credit recovery, and high school diploma program requirements, and thus integration would need to be coordinated across the options offered at those sites.

Formal Industry Engagement Was Going on in Two of the Five Sites Interviewed

The sites with formal relationships with local industry had developed these relationships organically over a long period of time. The sites with these partnerships were able to use them to provide both formal and informal employment opportunities, as well as to tailor some of the offerings to employer-expressed demand for certain skills.

Opportunities for and Challenges to Providing CTE Across ChalleNGe Sites

ChalleNGe program sites approached CTE in various ways to be able to provide it within the constraints of meeting their other requirements and implementing the eight core components. ChalleNGe sites were focused on ensuring that cadets are in a position to meet the requirements to fulfill a GED, credit recovery, and complete a high school diploma. This necessitated establishing criteria for eligibility to participate in CTE. For some CTE offerings, sites were able to offer courses to all cadets, but in other cases, where seats were limited, they generally factored in results on the TABE, progress on academic coursework, and behavioral indicators.

One of the biggest obstacles that sites reported in providing CTE had to do with scheduling constraints. It was often difficult for them to set aside time for CTE, particularly for sites that had to transport cadets to and from locations where they would be taking CTE classes. The time spent in transportation, as well as the logistics, limited the amount of CTE they could provide outside the ChalleNGe campus.

The purpose of the CTE analytic study was to delve more deeply into an important area that emerged during visits to ChalleNGe sites. Conversations with ChalleNGe leadership revealed that many saw the value of providing CTE to their cadets, particularly with the expansion of Job ChalleNGe on the horizon. It was also clear that provision of CTE was determined, to some extent, by the particular constraints that sites face. As more sites expressed interest in expanding CTE, it became clear that the RAND team could inform the decision-making by drawing on best-practices from the literature and facilitating some cross-sharing and pollination of ideas across sites. The full results of our study will be published separately as a stand-alone document. A follow-on to the CTE study could be to look closely at states with both ChalleNGe and Job ChalleNGe, and to consider ways in which offerings can be aligned to develop more-robust career pathways to better position ChalleNGe graduates to succeed in Job ChalleNGe and then to transition into well-paying jobs in subbaccalaureate fields.

Examining Mental Health Supports at ChalleNGe Sites

Mental health issues are an increasing concern among adolescents, and the youth served by ChalleNGe may be particularly at risk for mental health problems. Research suggests that about 20 percent of lower-income children have a diagnosed mental health condition (Burns et al., 2004), and almost 70 percent of youth in the juvenile justice system have a diagnosable mental health problem (Skowyra and Cocozza, 2006). In addition, almost 44 percent of youth with mental health problems drop out of high school (Wagner, 2005).

Indeed, interviews conducted during site visits to programs suggest that some cadets have previously diagnosed mental health conditions that require medication and/or counseling. According to our initial interviews, these conditions may or may not be disclosed by the youth or their caregivers during the application process, and the diagnosis can be revealed after the cadet has matriculated. Cadets can also have undiagnosed mental health conditions that emerge during their time at ChalleNGe.

Prior site visit interviews also suggest that some cadets have experienced traumatic events or been exposed to trauma in their families or communities (e.g., family or community violence) that could lead them to experience serious mental health concerns, such as posttraumatic stress disorder, anxiety, and depression. It is important to note that the experience of childhood trauma can have negative consequences on cadets' educational outcomes: Almost 20 percent of children who experience a traumatic event (e.g., sexual assault) during childhood drop out of high school, compared with 13 percent who did not experience childhood trauma (Porche et al., 2011).

Site visit interviews suggested that ChalleNGe programs have different staffing models, policies, and procedures for dealing with the mental health issues of cadets, from disqualifying applicants with mental health conditions to having licensed mental health professionals on staff to provide psychological services. Given the prevalence of mental health concerns among the ChalleNGe-eligible population, and the fact that, properly treated, mental health issues should not impede cadets from successful participation in the ChalleNGe program, the current study seeks to develop a set of recommendations for best-practices to address the mental health needs of cadets. To develop these recommendations, we will

1. review the literature on best-practices for identifying and treating adolescents with mental health concerns within the school setting
2. interview a set of counselors at ChalleNGe academies to better understand their policies and procedures for cadets with mental health concerns, their staffing model, what they are doing that is innovative in this area, and what they think would be best-practices for ChalleNGe programs
3. develop questions for the annual survey to assess the models currently in place across all ChalleNGe programs.

Results of this study will include a summary of the literature on best-practices for mental health practices in a school setting, a summary of the current state of mental health counseling models at different academies, and case studies of innovative practices or programs used by academies. These results will be used to make recommendations for ChalleNGe program staffing, policies, and procedures to help programs support the mental health needs of cadets. The goal of these recommendations will be to provide programs with a set of best-practices to maximize the ability of all cadets to successfully complete the program and become productive citizens.

Study of High-Paying Job Skills

A separate analytic product that we are producing for the benefit of site directors and ChalleNGe national leadership is a study of high-paying job skills. We are motivated by the finding that many ChalleNGe sites are seeking opportunities to expand job training for their cadets. However, on-site training can be expensive or infeasible given facility constraints, while off-site training requires building partnerships and may accommodate only a limited number of cadets. The aim of this analysis, which we refer to as the *Occupation Report*, is to identify the skills prevalent among occupations that are high-paying, high-growing, or both, so that sites can incorporate skill-training, if not job-training, into their curriculum.

The Occupation Report will identify "good jobs" for workers without a college degree. We use the Current Population Survey, a nationally representative household survey that is used to calculate most national employment and unemployment statistics. We start by classifying all occupations in the Current Population Survey by the average education of the workers within those occupations. Our aim is to identify occupations that do not require a four-year or advanced degree. Within those noncollege occupations, we then define a *good job* as an occupation that is growing in number and whose workers earn above the median for their education, race, and gender group. In other words, we look for occupations that site directors would encourage cadets to plan for—within reach given their current or planned education, but also sustainable and high-paying. Once we have our set of occupations, we then use the O*NET, a national database of comprehensive occupation descriptions that is frequently used in job interest surveys, to look for common and highly ranked skills, abilities, and job features among the good jobs.

For example, we may learn that many occupations that we identify as good jobs put a high emphasis on skills, abilities, and features that involve working on a team: "communicating with supervisors, peers, or subordinates," "establishing and maintaining interpersonal relationships," "oral expression," "resolving conflicts and negotiating with others," and "guiding, directing, and motivating subordinates." This information can inform ChalleNGe sites as they establish job training and life skills curricula by providing examples of what we know is associated with labor market rewards. The Occupation Report will present skill summaries of good jobs by levels of education (such as high school diploma holder versus occupational certificate holder), and by industry (such as construction versus sales).

In addition to the analytic efforts described above, the RAND team is conducting several other ongoing efforts that were informed through the visits to ChalleNGe sites. One such effort is the examination of mental health supports as more ChalleNGe sites become aware of the mental and behavioral health needs of their incoming youth. The other is a study of high-paying job skills.

Description of Pilot Projects

In addition to the analytic efforts, RAND has also assisted two program sites with implementing two distinct pilots. In one case, the pilot drew on information from a review of the mentoring literature published in a previous report (Constant et al., 2019) to design a mentor training pilot suitable to a ChalleNGe site. Below, we describe the design of the mentoring pilot and share early findings from implementing it in one of the ChalleNGe sites. The other pilot was working with a ChalleNGe site to implement a survey of graduates or alumni well past the Post-Residential Phase. This was to address the difficulty that sites faced in gathering information on graduates past the Post-Residential Phase.

Improving Mentor Training

Mentoring is a critical component of ChalleNGe. Mentors support cadets in achieving their goals and staying on the right path during the Post-Residential Phase. In addition, mentors are responsible for reporting on cadets' placement (e.g., employment) during the Post-Residential

Phase. In Constant et al. (2019), we reviewed the literature on mentoring and analyzed site visit interviews to better understand ChalleNGe mentoring practices. The analysis revealed that there was a need to further develop mentors' engagement with mentees. All sites provided training to mentors, and the most typical training is a one-day training that takes place onsite. In some cases, sites provided training that lasted over multiple days. However, the analysis revealed that there were challenges in the mentor-mentee relationship, with communication and engagement diminishing over time. Not only did some sites report that mentors were overwhelmed with the responsibilities of reporting placement information back to the sites, but sites reported that mentors struggled to keep their mentees engaged, and oftentimes the relationship did not meet joint expectations. In fact, sites reported significant problems with retaining mentors (as measured by the scheduled reporting of placement information) throughout the Post-Residential Phase, with loss of contact ranging from 20 to 80 percent of mentors over the Post-Residential period (Constant et al., 2019).

The concerns that sites reported with the mentor-mentee relationships prompted the RAND team to explore ways to improve this component of ChalleNGe. The RAND team partnered with a ChalleNGe site to identify gaps in the site's mentoring practices, strengthen mentor training, and propose ways to monitor changes to mentor-mentee engagement during the Post-Residential Phase. The initial findings from that pilot are documented in this report and described below. A more comprehensive assessment of the pilot through multiple classes will be presented in the final report for this study.

A Partnership with Sunburst Youth Academy

The RAND team regularly engages with different ChalleNGe program sites to learn about their needs and to collaborate on analytic tasks. Sunburst Youth Academy, based in Los Alamitos, California, expressed an interest in improving its mentoring program during a site visit. The RAND team worked collaboratively with Sunburst to design and implement a pilot project to improve its mentoring component. Members of the RAND study team attended Sunburst's mentor orientation in January 2019 to gain a deeper understanding of the current mentor training program. The mentor orientation focused on introducing the Youth ChalleNGe program (e.g., review of the eight core components) and discussing the mentor's responsibilities. Following the orientation, RAND met with Sunburst's senior case manager, outreach supervisor, and director to discuss options to enhance subsequently scheduled mentor trainings by integrating a new training module that targets the mentor-mentee relationship.

Gap Analysis of Mentor Training

Although research has repeatedly identified the importance of mentor training (DuBois et al., 2011), evidence-based mentor training curricula have not been developed and made available. However, MENTOR: The National Mentoring Partnership has developed guidance on Elements of Effective Practice for Mentoring, and one of the effective practice elements is mentor training (Garringer et al., 2015).[4] For each element, the authors recommended a set of benchmarks based on evidence from both research and practice. One benchmark for mentor training is to include pre-match training for mentors on the following topics: program requirements, mentors' goals and expectations, mentors' obligations and responsibilities, relationship devel-

---

[4] MENTOR is a national mentor advocacy organization whose mission it is to "fuel the quantity and quality of mentoring relationships for America's young people to close the mentoring gap." For more information, see MENTOR (undated).

opment and maintenance, ethical and safety issues, initiation and closure of the mentoring relationship, developmental issues (e.g., adolescents' need for autonomy), and population-specific considerations (e.g., cultural background of mentee). Garringer and colleagues, however, did not recommend any training materials on any of these topics.

RAND reviewed Sunburst's mentoring training materials carefully and identified gaps in some training topics. First, we reviewed Sunburst's training materials to identify topics that are covered and compared these topics with the ones recommended by MENTOR's Elements of Effective Practice for Mentoring. We found that Sunburst's training includes very limited information on a few recommended topics: relationship development and maintenance, initiation and closure of the mentoring relationship, developmental issues, and population-specific considerations. The topic of relationship development and maintenance is the most relevant to addressing Sunburst's concerns about mentor retention and the mentor-mentee relationship. Then, RAND compared Sunburst's mentor training materials on relationship development to a few mentor training curricula recommended by the National Mentoring Resource Center to identify specific training modules to include in the pilot.[5] We found that Sunburst's training did not include any information on effective communication skills. RAND shared the review with Sunburst and suggested the addition of a training module on mentor communication skills to improve mentor-mentee relationships. The recommended training module on communication skills was adapted from the *Training New Mentors* guide published by the Hamilton Fish Institute on School and Community Violence and the National Mentoring Center at Northwest Regional Educational Laboratory (2007).

On March 9 and March 10, 2019, Sunburst delivered the mentor training with the additional module on communication skills. The module covered two topics: active listening and empathy, and included three activities (see Table 3.3). The first activity took 25 minutes, and the second and third activities each took 40 minutes to complete. Each activity included role playing, group discussions, and handouts.

Feedback from Mentors

The RAND team conducted a posttraining assessment to solicit feedback from mentors on the mentor training in general and specifically on the newly proposed communication skills module. RAND and Sunburst developed a survey that included the following four questions: (1) How helpful was the training? (2) How helpful were the active listening skills discussed during the training? (3) What was the most useful part of the training? and (4) What additional topics would you like the training to cover?

One hundred fifty mentors participated in the training and completed the short survey. The majority of mentors were male (103 of 150), and most of them were family friends of cadets (82 of 150). On average, mentors perceived both the overall training and specific communication skills module to be helpful or very helpful. The average score for each of the first two questions was 2.6 (zero = not at all helpful, 1 = somewhat helpful, 2 = helpful, and 3 = very helpful). No respondents rated the training to be "not at all helpful." When asked what part of the training was most useful, the mentors reported the following:

- tips on communication, listening, and interacting with youth
- role of mentor and role of cadet

[5] For more information, visit its website at nationalmentoringresourcecenter.org.

- meeting with the cadets and interacting with them in the context of the training
- listening to the cadets talk about their experience
- program orientation and expectations of mentor
- reviewing and signing the mentor-cadet agreement.

Participants suggested the following additional topics for future trainings (Figure 3.2): career exploration (69 percent); family relationships (54 percent); and addressing behavioral issues (47 percent), emotional issues (46 percent), and academic issues (46 percent).

Next Steps for the Mentoring Pilot

RAND is continuing the partnership with Sunburst Youth Academy to collect follow-up data from mentors during the Post-Residential Phase. Mentors will complete an online survey every quarter during this period to report on the frequency of contact with the mentee, forms of communication (e.g., in-person meeting, phone calls, emails), mentor's confidence in supporting the mentee, frequency of applying active listening skills when communicating with the

Table 3.3
Module on Communication Skills Training for Mentors

Activity	Objectives	Steps
First—"If you want easy listening, turn on the radio"	1. To learn the difference between supportive and nonsupportive communication 2. To understand the qualities of active listening	1. Facilitators do the first role play of a conversation between a mentor and a mentee to display poor listening skills, then ask participants for feedback about how to improve listening skills. 2. Facilitators do the second role play to display good listening skills then ask participants for feedback again. 3. Distribute "I Hear You" handout and encourage participants to ask questions.
Second—"Communication role plays"	1. To practice applying active listening skills 2. To practice applying supportive listening skills	1. Divide participants into pairs. Each pair practices a scenario from the handout. 2. Participants should practice using both poor and good listening skills. 3. Bring the group together and ask one pair to volunteer to role play their conversation and ask group members to give feedback and suggestions.
Third—"Communication role plays with mentees"	1. To practice applying active listening skills with mentees 2. To practice applying supportive listening skills with mentees	1. Pair mentor to his/her mentee. Each pair practices a scenario from the handout. 2. Mentors should practice using both poor and good listening skills. 3. Bring the group together and ask one pair to volunteer to role play their conversation and ask group members to give feedback and suggestions.

SOURCE: Hamilton Fish Institute on School and Community Violence and The National Mentoring Center at Northwest Regional Educational Laboratory (2007).

Figure 3.2
Additional Topics Requested by Mentors to Be Covered in Future Trainings

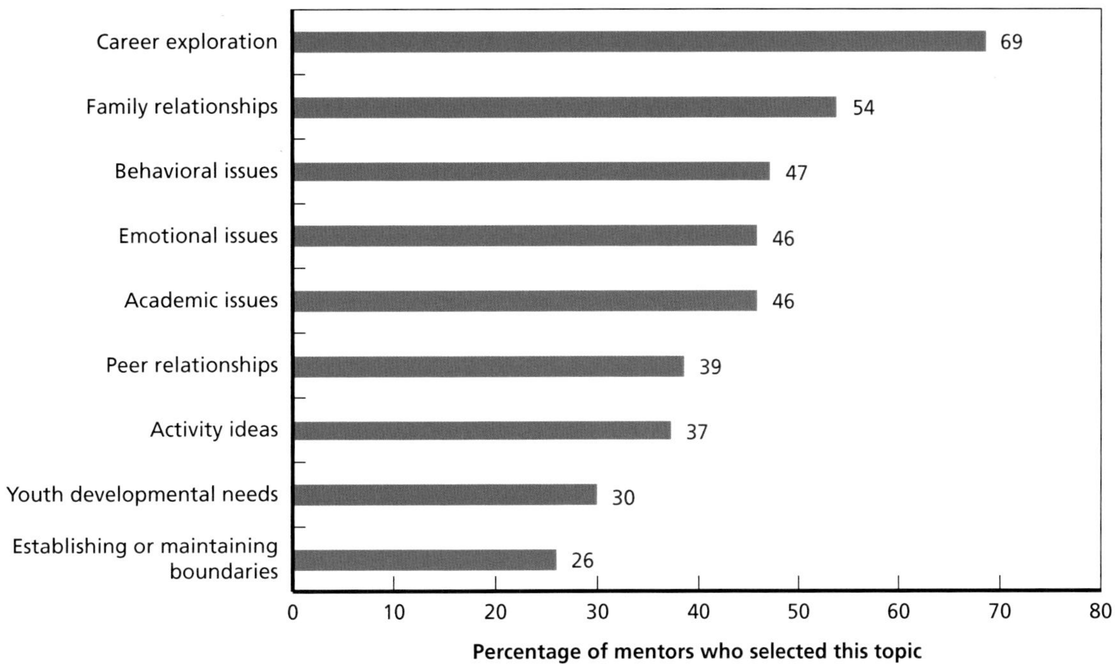

Career exploration 69
Family relationships 54
Behavioral issues 47
Emotional issues 46
Academic issues 46
Peer relationships 39
Activity ideas 37
Youth developmental needs 30
Establishing or maintaining boundaries 26
0 10 20 30 40 50 60 70 80
Percentage of mentors who selected this topic

SOURCE: Mentor posttraining survey.

mentee, challenges related to keeping in touch with the mentee, and the quality of the mentor-mentee relationship. The goal of this survey is twofold: (1) to track the application of active listening skills (i.e., the added communication skills module); and (2) to examine changes in the mentor-mentee relationship over time.

Furthermore, RAND and Sunburst Youth Academy are currently testing the mentoring pilot with a second cohort of mentors. The same communication skills module was delivered to mentors on September 14 and September 15, 2019. We are currently collecting posttraining data from mentors to seek feedback about the training. One additional training will be tested with the second cohort—Growth Mindset for Mentors, which was developed by the Project for Education Research That Scales and MENTOR.[6] *Growth mindset* is a catchphrase for the belief that intelligence can be developed through effort, in contrast to a belief that intelligence is fixed (Dweck, 2006). The online training aims to help mentors understand growth mindset and to learn growth mindset strategies that can help them support their mentees. Mentors are expected to complete the training before graduation. RAND and Sunburst plan to follow up with this cohort of mentors during the Post-Residential Phase as well. A quarterly survey will be distributed to mentors online during the Post-Residential Phase to track the application of active listening skills and growth mindset strategies and to examine the evolution of the quality of mentor-mentee relationship over time.

[6] See Project for Education Research That Scales and MENTOR (undated).

Collecting Long-Term Outcome Information from an Alumni Survey

The RAND research team is working with the Washington Youth Academy to identify means of tracking cadets postgraduation using a simple survey to collect placement data, as well as information on longer-term outcomes, such as education, employment, and life transitions. Site visits reveal that all ChalleNGe sites face difficulties in tracking cadets well into the Post-Residential Phase, let alone beyond that phase. Much of the information gathered on cadet well-being is anecdotal and not representative of the variety of experiences graduates are likely to have. The RAND research team worked with the site to determine the feasibility of surveying program alumni; the site advertised and then implemented a short survey to collect information on ChalleNGe alumni experiences and outcomes, including educational attainment, labor market experiences, and family formation. While the response rate was relatively modest, at around 16 percent,[7] initial results appear very promising—the survey responses include information on a variety of longer-term outcomes, such as postsecondary educational attainment, earnings, job stability, and family formation. The results appear sensible (for example, earnings are higher among those who are older and have more education). There is substantial variation among respondents in terms of age, gender, ethnicity, and region; using these measures, the RAND team is working to develop a series of weights to produce results that would be representative of all alumni of that program.

Summary

In this chapter, we reviewed our logic model. We also described the schedule of ChalleNGe site visits, including the methodology that we are using to collect and analyze data from the sites. Finally, we described our analytic efforts and the two pilots that we are working on with ChalleNGe sites. All these support the operation of the ChalleNGe program and will assist in determining the extent to which the program is meeting its mission. Taken as a group, the analytic efforts and pilots address many of the core components of ChalleNGe and will produce actionable recommendations to improve program outcomes. In Chapter Four, we offer some concluding thoughts.

---

[7] The survey was administered in November 2018. As of then, there were 2,323 graduates of the program. The number of graduates who started the survey was 421, and the number who completed it was 374.

CHAPTER FOUR

Concluding Thoughts

The National Guard Youth ChalleNGe program continues to provide opportunities for young people who struggle in traditional high schools. Cadets who participate in ChalleNGe have the opportunity to earn a combination of academic credentials—certificates (such as the GED or HiSET), high school credits that allow them to re-enroll in their home high schools, or high school diplomas. Along with academic credentials, cadets can gain a variety of occupational experience and training, perform community service, improve their physical fitness, and develop their noncognitive or socioemotional skills. The focus on the eight core components of ChalleNGe ensures that cadets gain a variety of experiences and skills.

To date, more than 234,000 young people have participated in ChalleNGe, and close to 175,000 have completed the program. ChalleNGe programs took in fewer participants in 2018 than in 2017, and this decrease can be attributed to a decline in the size of the programs as well as a more challenging recruiting environment than existed previously. Unlike previous years, no new programs were added between 2017 and 2018. Among cadets who entered ChalleNGe in 2018, a little more than 9,300 graduated from one of the 39 sites included in our data collection, and more than 4,000 received at least one education credential. The overall graduation rate has remained roughly constant across the years.

Data collected from the sites demonstrate that ChalleNGe cadets show marked improvement in academic skills (measured by the proportion who achieve key benchmarks on the TABE), though the continued adoption of the new TABE (11/12) will result in lower scores for ChalleNGe cadets overall. Scores from the TABE 11/12 have not been aligned with current cadet outcome measures, and programs should interpret the results from the TABE 11/12 carefully and not assume that the lower scores necessarily imply a decline in achievement.

The study also examined ChalleNGe program staffing with a specific focus on staff turnover. Staff turnover, particularly among the cadre staff, is high across programs, though in some programs more than others. Though they are descriptive and suggestive in nature, data from the annual survey revealed that programs where half or more of cadre were hired in the past 12 months had significantly lower starting salaries for cadre. Similarly, program sites where one-third or more of their instructional staff were hired in the past 12 months reported lower starting salaries for instructional staff than other sites. Turnover can be disruptive to sites and, especially, to cadets, and while salary is not the only factor that determines turnover, it is one aspect of program policy and practice that may be worth further exploring.

Since the inception of the RAND team's current ChalleNGe project, we have collected four rounds of data from the ChalleNGe sites and have carried out 39 site visits. This report is the fourth that supports ChalleNGe's reporting requirement to Congress. In summer 2020, we will produce a capstone report that summarizes the data we have collected to date and lays out a series of recommendations to measure cadets' long-term outcomes. That report will include information on a variety of different methods for measuring long-term outcomes, with the ultimate goal of determining how well the ChalleNGe program is doing in meeting its mission "to intervene in and reclaim the lives of 16–18-year-old high school dropouts, producing program graduates with the values, life skills, education, and self-discipline necessary to succeed as productive citizens."

In addition to collecting annual program- and cadet-level indicators, the RAND research team is carrying out several analytic efforts. These efforts include research on a variety of topics that are relevant to the core components, including mentorship, CTE, mental health counseling, and high-paying job skills. The research team will also continue to work on developing program-level indicators and exploring associations with outcomes of interest, including performance on TABE, graduation, and postresidential placement. In this report, we described our partnership with a ChalleNGe academy to pilot a mentor training program, and we shared findings from examining CTE implementation across ChalleNGe sites.

The mentor training pilot is ongoing, and results will be shared in future reports. The training incorporated best-practices from the literature and was designed to address a key problem faced by ChalleNGe sites—maintaining mentor and mentee engagement in the Post-Residential Phase. The pilot includes plans for collecting data on mentors to monitor progress in the use of communication strategies introduced in the training. The outcome of this pilot will be shared in a future report, with lessons learned that are potentially useful across ChalleNGe sites.

The study of CTE practices across ChalleNGe sites highlighted the difficulties with integrating occupationally focused learning into a compressed schedule. The study identified promising practices from the evidence base, and some ChalleNGe programs are finding ways to incorporate some of those practices. Though ChalleNGe sites admit that more needs to be done, finding creative ways to provide quality CTE within the existing constraints will remain a focus. The expansion of Job ChalleNGe from the initial three pilot states reinforces the need to prepare cadets in ChalleNGe to succeed in their postsecondary transitions, be it further education, training, or employment. The mental health and jobs study, the findings of which will be published in future reports, will also further inform ChalleNGe programs to make improvements and better prepare cadets to succeed both at ChalleNGe and beyond.

APPENDIX A

Site-Specific Information

This appendix includes a complete list of the ChalleNGe programs, as well as the program-level tables of information. Table A.1 provides the complete name and location (state) of each program.

Table A.1
National Guard Youth ChalleNGe: Program Abbreviation, State, and Name

Program Abbreviation	State	Program Name	Program Type
AK	Alaska	Alaska Military Youth Academy	High school credits or diploma, GED
AR	Arkansas	Arkansas Youth ChalleNGe	GED
CA-DC	California	Discovery ChalleNGe Academy	High school credits or diploma, GED
CA-LA	California	Sunburst Youth Academy	High school credits or diploma, HiSET
CA-SL	California	Grizzly Youth Academy	High school credits or diploma, HiSET
D.C.	District of Columbia	Capital Guardian Youth ChalleNGe Academy	GED
FL	Florida	Florida Youth ChalleNGe Academy	High school credits or diploma, GED
GA-FG	Georgia	Fort Gordon Youth ChalleNGe Academy	High school credits or diploma, GED
GA-FS	Georgia	Fort Stewart Youth ChalleNGe Academy	High school credits or diploma, GED
GA-MV	Georgia	Milledgeville Youth ChalleNGe Academy	High school credits or diploma, GED
HI-BP	Hawaii	Hawaii Youth ChalleNGe Academy at Barbers Point	High school credits or diploma, HiSET
HI-HI	Hawaii	Hawaii Youth ChalleNGe Academy at Hilo	High school credits or diploma, HiSET
ID	Idaho	Idaho Youth ChalleNGe Academy	High school credits or diploma, GED
IL	Illinois	Lincoln's ChalleNGe Academy	High school credits or diploma, GED

Table A.1—Continued

Program Abbreviation	State	Program Name	Program Type
IN	Indiana	Hoosier Youth ChalleNGe Academy	TASC
KY-FK	Kentucky	Bluegrass ChalleNGe Academy	High school credits or diploma, GED
KY-HN	Kentucky	Appalachian ChalleNGe Program	High school credits or diploma, GED
LA-CB	Louisiana	Louisiana Youth ChalleNGe Program–Camp Beauregard	High school credits or diploma, HiSET
LA-CM	Louisiana	Louisiana Youth ChalleNGe Program–Camp Minden	High school credits or diploma, HiSET
LA-GL	Louisiana	Louisiana Youth ChalleNGe Program–Gillis Long	High school credits or diploma, HiSET
MD	Maryland	Freestate ChalleNGe Academy	High school credits or diploma
MI	Michigan	Michigan Youth ChalleNGe Academy	High school credits or diploma, GED
MS	Mississippi	Mississippi Youth ChalleNGe Academy	High school credits or diploma
MT	Montana	Montana Youth ChalleNGe Academy	High school credits or diploma, HiSET
NC-NL	North Carolina	Tarheel ChalleNGe Academy–New London	High school credits or diploma, HiSET, GED
NC-S	North Carolina	Tarheel ChalleNGe Academy–Salemburg	High school credits or diploma, HiSET, GED
NJ	New Jersey	New Jersey Youth ChalleNGe Academy	GED
NM	New Mexico	New Mexico Youth ChalleNGe Academy	HiSET
OK	Oklahoma	Thunderbird Youth Academy	High school credits or diploma, GED
OR	Oregon	Oregon Youth ChalleNGe Program	High school credits or diploma, GED
PR	Puerto Rico	Puerto Rico Youth ChalleNGe Academy	High school credits or diploma
SC	South Carolina	South Carolina Youth ChalleNGe Academy	GED
TN	Tennessee	Volunteer Youth ChalleNGe Academy	High school credits or diploma, HiSET
TX	Texas	Texas ChalleNGe Academy	High school credits or diploma, GED
VA	Virginia	Virginia Commonwealth ChalleNGe Youth Academy	High school credits or diploma, GED
WA	Washington	Washington Youth Academy	High school credits or diploma

Table A.1—Continued

Program Abbreviation	State	Program Name	Program Type
WI	Wisconsin	Wisconsin ChalleNGe Academy	High school credits or diploma, GED
WV	West Virginia	Mountaineer ChalleNGe Academy	High school credits or diploma
WY	Wyoming	Wyoming Cowboy ChalleNGe Academy	High school credits or diploma, HiSET

The following tables include detailed information collected from each program. We carried out data collection in July and August 2018. The focus of the data collection was on classes that began in 2018 (Classes 50 and 51, according the ChalleNGe class numbering system, which began with the first class in the 1990s).

In some cases, programs provided incomplete data or data that were suspect in some way. When this occurred, we indicated the specific elements that were not reported. Some of these data issues are related to the variation in how the individual sites collect and store data. RAND analysts are currently exploring strategies to increase the accuracy of future data collected from the sites, with a focus on limiting the burden of data collection for sites and ChalleNGe personnel.

The sites are listed alphabetically by state or territory abbreviation. Each table includes metrics of the number and type of staff, total funding in 2018, as well as the numbers of cadets who applied, entered, graduated, and received various credentials. The tables also include data related to several of the core components—service to community (and calculated values based on local labor market conditions), gains on specific physical fitness tests, as well as the numbers of cadets registered to vote and for Selective Service. Finally, the tables include information about postgraduation placement (although there is no information on Classes 50 and 51's 12-month placement rates because fewer than 12 months have passed since graduation for this group). The tables also include 12-month placement rates for Class 49; at the time of our previous data collection, 12-month information was not yet available for cadets in Class 49.

Some of the data in the following tables (along with other cadet-level data collected at the same time) formed the basis of analyses presented in Chapter Two. These same data will also be used in some of our future analyses, which we described in Chapter Three.

Table A.2
Applicants and Graduates (Classes 50 and 51)

	Residential Class 50				Residential Class 51			
Site	Target	Applied	Entrants	Graduates	Target	Applied	Entrants	Graduates
All sites	*	9,578	6,445	4,644	*	9,679	6,399	4,707
AK	144	248	227	187	144	221	195	156
AR	100	216	135	80	100	257	162	107
CA-DC	125	200	133	120	125	259	146	129
CA-LA	190	307	196	147	190	420	209	186
CA-SL	190	311	235	184	190	336	222	192
D.C.	79	158	50	31	75	180	68	41

Table A.2—Continued

	Residential Class 50				Residential Class 51			
Site	Target	Applied	Entrants	Graduates	Target	Applied	Entrants	Graduates
FL	150	174	174	137	150	214	207	159
GA-FG	*	306	252	182	*	223	186	140
GA-FS	212	312	218	174	213	335	260	199
GA-MV	150	182	142	86	150	189	189	102
HI-BP	125	130	105	94	125	122	101	79
HI-HI	75	85	74	68	75	127	84	63
ID	105	166	129	111	105	170	150	129
IL	175	284	236	116	175	272	205	123
IN	100	145	122	72	100	129	93	61
KY-FK	100	140	116	65	100	83	65	41
KY-HN	100	149	119	74	100	111	90	90
LA-CB	250	424	276	204	250	427	296	201
LA-CM	200	332	254	204	200	319	233	174
LA-GL	250	459	333	229	250	379	276	188
MD	100	206	148	71	100	273	163	98
MI	114	196	143	115	114	251	154	118
MS	200	456	291	192	200	495	286	201
MT	100	182	147	113	100	153	133	102
NC-NL	100	340	156	113	100	344	148	109
NC-S	125	388	163	103	125	333	151	110
NJ	100	283	128	74	100	269	126	80
NM	100	174	142	112	100	139	110	85
OK	110	455	164	103	110	396	181	105
OR	125	232	156	136	125	232	157	139
PR	200	295	261	217	220	301	261	227
SC	100	96	93	60	100	171	144	102
TN	100	104	65	46	100	144	98	53
TX	100	189	102	63	100	216	155	87
VA	125	233	167	104	125	213	135	91
WA	135	264	166	137	135	304	165	145
WI	100	246	140	100	100	221	126	88

Table A.2—Continued

	Residential Class 50				Residential Class 51			
Site	Target	Applied	Entrants	Graduates	Target	Applied	Entrants	Graduates
WV	150	358	174	139	150	393	219	173
WY	60	153	113	81	47	58	50	34

NOTES: Information in this table was reported by the sites in July and August 2019 and covers Classes 50 and 51, which began in 2018. Target columns represent the program's graduation goal. Additional information on each ChalleNGe site is available throughout Appendix A.

* = did not report information.

Table A.3
Number of ChalleNGe Graduates and Number of Graduates by Type of Credential Awarded, by Site (Classes 50 and 51)

	Residential Class 50				Residential Class 51			
Site	Number of Graduates from ChalleNGe	Number Receiving GED, HiSET, or TASC	Number Receiving HS Credits	Number Receiving HS Diploma	Number of Graduates from ChalleNGe	Number Receiving GED, HiSET, or TASC	Number Receiving HS Credits	Number Receiving HS Diploma
AK	187	26	137	1	156	22	107	2
AR	80	21	0	0	107	44	0	0
CA-DC	120	0	86	34	129	0	104	25
CA-LA	147	0	121	26	186	0	161	25
CA-SL	184	8	123	53	192	4	122	66
D.C.	31	14	0	0	41	12	0	0
FL	137	83	16	2	159	116	24	8
GA-FG	182	57	0	36	140	36	33	15
GA-FS	174	39	24	49	199	20	28	62
GA-MV	86	29	14	9	102	30	28	18
HI-BP	94	0	0	94	79	15	0	64
HI-HI	68	0	0	64	63	0	0	61
ID	111	0	97	14	129	0	112	17
IL	116	65	0	0	123	67	0	0
IN	72	53	0	0	61	42	0	0
KY-FK	65	0	49	12	41	0	35	6
KY-HN	74	0	73	0	90	0	89	0

Table A.3—Continued

	Residential Class 50				Residential Class 51			
Site	Number of Graduates from ChalleNGe	Number Receiving GED, HiSET or TASC	Number Receiving HS Credits	Number Receiving HS Diploma	Number of Graduates from ChalleNGe	Number Receiving GED, HiSET or TASC	Number Receiving HS Credits	Number Receiving HS Diploma
LA-CB	204	65	0	0	201	68	0	0
LA-CM	204	58	7	1	174	53	0	0
LA-GL	229	82	0	0	188	71	0	0
MD	71	0	0	40	98	0	0	65
MI	115	0	89	26	118	0	75	43
MS	192	0	0	119	201	0	0	123
MT	113	57	0	0	102	52	0	0
NC-NL	113	44	0	24	109	53	0	26
NC-S	103	49	0	27	110	62	0	11
NJ	74	4	0	0	80	19	0	0
NM	112	66	0	0	85	53	0	0
OK	103	11	81	11	105	5	92	8
OR	136	0	124	12	139	0	129	10
PR	217	0	0	215	227	0	0	226
SC	60	24	0	0	102	20	0	0
TN	46	*	*	*	53	*	*	*
TX	63	2	50	8	87	4	66	11
VA	104	45	42	0	91	31	55	0
WA	137	0	137	0	145	0	145	0
WI	100	41	0	59	88	20	0	68
WV	139	0	21	118	173	0	17	156
WY	81	38	0	0	34	22	0	0

NOTES: HS = high school. Information in this table was reported by the sites in July and August 2019 and covers Classes 50 and 51, which began in 2018. Credentials awarded include those conveyed during the course of the ChalleNGe Residential Phase. Counts reflect a single credential per cadet. Cadets with multiple credentials are assigned based on the hierarchy of HS diploma, HS credits, then GED/HiSET/TASC. At the Idaho ChalleNGe program, those who received GEDs also received high school credits, although the credits were not used. In New Jersey, ChalleNGe graduates who pass the GED are awarded a state high school diploma. In West Virginia, ChalleNGe graduates who pass the state standardized test are awarded a state high school diploma. The Wisconsin program generates a pathway for all credentialing options awarded through the Wisconsin Department of Instruction and associated school districts, including credit recovery, GED, a high school equivalency diploma, and a high school diploma. Additional information on each ChalleNGe site is available throughout Appendix A.

* = did not report.

Table A.4
Core Component Completion—Community Service, ChalleNGe Graduates (Classes 50 and 51)

Site	Residential Class 50			Residential Class 51		
	Service Hours/ Cadet	Dollar Value/ Hour	Total Community Service Contribution	Service Hours/Cadet	Dollar Value/ Hour	Total Community Service Contribution
All sites	N/A	N/A	$7,014,366	N/A	N/A	$7,032,294
AK	62	$27.88	$322,761	49	$27.88	$212,754
AR	98	$20.49	$160,437	71	$20.49	$154,945
CA-DC	44	$29.95	$158,136	52	$29.95	$200,905
CA-LA	42	$29.95	$184,911	43	$29.95	$239,540
CA-SL	48	$29.95	$264,518	48	$29.95	$276,019
D.C.	64	$41.72	$82,190	52	$41.72	$89,323
FL	83	$24.04	$271,712	79	$24.04	$300,055
GA-FG	56	$25.78	$260,573	43	$25.78	$154,737
GA-FS	52	$25.78	$233,257	73	$25.78	$372,624
GA-MV	43	$25.78	$98,572	44	$25.78	$116,174
HI-BP	156	$26.87	$393,969	121	$26.87	$256,394
HI-HI	117	$26.87	$213,778	116	$26.87	$199,483
ID	48	$22.14	$116,979	48	$22.14	$136,805
IL	51	$26.89	$159,050	74	$26.89	$246,609
IN	56	$24.13	$96,771	52	$24.13	$75,952
KY-FK	65	$21.42	$89,135	75	$21.42	$65,849
KY-HN	57	$24.19	$101,407	44	$24.19	$71,717
LA-CB	50	$22.76	$230,377	52	$22.76	$238,183
LA-CM	50	$22.76	$235,016	66	$22.76	$263,276
LA-GL	46	$22.76	$239,367	69	$22.76	$294,753
MD	54	$28.65	$110,174	46	$28.65	$129,627
MI	57	$24.85	$161,637	64	$24.85	$188,279
MS	65	$19.70	$245,856	72	$19.70	$285,098
MT	58	$23.09	$150,836	62	$23.09	$146,563
NC-NL	87	$24.19	$236,358	91	$24.19	$239,545
NC-S	81	$24.19	$201,169	67	$24.19	$178,174
NJ	48	$28.82	$102,369	42	$28.82	$96,835
NM	61	$21.20	$144,838	54	$21.20	$97,308
OK	76	$22.95	$179,348	77	$22.95	$185,666
OR	88	$25.40	$303,642	94	$25.40	$333,289

Table A.4—Continued

Site	Residential Class 50			Residential Class 51		
	Service Hours/Cadet	Dollar Value/Hr.	Total Community Service Contribution	Service Hours/Cadet	Dollar Value/Hr.	Total Community Service Contribution
PR	40	$12.64	$109,715	40	$12.64	$114,771
SC	46	$23.21	$64,060	28	$23.21	$66,288
TN	54	$22.67	$50,191	41	$22.67	$50,191
TX	64	$25.10	$100,618	51	$25.10	$109,117
VA	105	$27.50	$300,300	58	$27.50	$145,145
WA	56	$31.72	$241,897	65	$31.72	$300,341
WI	59	$25.12	$148,208	78	$25.12	$172,424
WV	48	$22.29	$149,800	49	$22.29	$190,314
WY	47	$24.60	$100,434	43	$24.60	$37,220

NOTES: Information in this table was reported by the sites in July and August 2019 and covers Classes 50 and 51, which began in 2018. The value of community service is calculated using published figures at the state level for 2018 and that are available online (Independent Sector, undated). The value of community service was calculated in the same manner in the previous annual reports (Constant et al., 2019; Wenger, Constant, and Cottrell, 2018; Wenger et al., 2017; National Guard Youth ChalleNGe, 2015).

N/A = not available.

Table A.5
Residential Performance—Physical Fitness as Measured by the Average Number of Initial and Final Push-ups Completed and Initial and Final Run-Times for Graduates, per Site (Class 50)

Site	Push-Ups		1-Mile Run	
	Initial	Final	Initial	Final
All sites	25	42	10:19	08:29
AK	27	44	10:41	08:49
AR	29	46	10:31	09:22
CA-DC	26	44	09:16	07:24
CA-LA	26	47	10:12	08:04
CA-SL	21	38	10:05	07:30
D.C.	23	41	10:19	09:44
FL	18	34	10:24	08:22
GA-FG	27	39	09:45	09:14
GA-FS	31	45	10:46	09:06
GA-MV	43	49	09:41	08:13
HI-BP	27	47	10:29	07:52

Table A.5—Continued

Site	Push-Ups		1-Mile Run	
	Initial	Final	Initial	Final
HI-HI	33	41	09:29	09:09
ID	22	44	09:35	07:37
IL	21	41	11:20	08:56
IN	22	42	10:27	08:15
KY-FK	29	38	12:32	10:39
KY-HN	30	52	09:34	08:55
LA-CB	25	44	09:34	08:08
LA-CM	18	34	10:42	07:53
LA-GL	29	43	10:16	09:32
MD	26	40	11:39	10:11
MI	30	47	09:20	07:55
MS	19	45	11:42	07:56
MT	21	39	11:04	08:12
NC-NL	21	36	12:38	10:27
NC-S	21	36	11:29	08:14
NJ	29	39	11:38	09:46
NM	28	47	08:12	06:38
OK	26	43	11:16	10:03
OR	27	51	09:07	07:24
PR	24	40	09:38	08:14
SC	28	43	10:13	09:01
TN	*	*	*	*
TX[a]	7	12	11:21	10:32
VA	22	44	08:54	08:23
WA	17	42	10:32	07:04
WI	14	23	11:19	07:18
WV	*	*	10:31	08:17
WY	24	37	08:21	07:41

NOTE: Information in this table was reported by the sites in July and August 2019 and covers Class 50.

* = did not report.

[a]Pull-ups; site does not collect data on push-ups.

Table A.6
Residential Performance—Physical Fitness as Measured by the Average Number of Initial and Final Push-ups Completed and Initial and Final Run-Times for Graduates, per Site (Class 51)

	Push-Ups		1-Mile Run	
Site	Initial	Final	Initial	Final
All sites	25	42	10:19	08:37
AK	25	42	10:10	09:01
AR	31	47	10:26	10:05
CA-DC	29	47	09:35	07:41
CA-LA	26	49	09:28	07:40
CA-SL	20	36	09:23	07:41
D.C.	26	39	12:05	11:53
FL	15	29	11:19	09:01
GA-FG	32	36	10:15	09:34
GA-FS	31	47	10:17	09:02
GA-MV	*	37	09:51	08:41
HI-BP	32	53	10:51	08:09
HI-HI	34	48	08:51	09:05
ID	27	40	10:08	07:43
IL	27	43	11:53	09:09
IN	32	46	12:13	09:49
KY-FK	23	47	10:14	09:06
KY-HN	23	47	09:51	08:34
LA-CB	32	36	09:56	09:27
LA-CM	21	34	12:30	08:31
LA-GL	27	41	10:51	09:23
MD	19	52	09:38	07:57
MI	28	47	08:50	07:57
MS	24	50	11:05	07:51
MT	22	39	10:51	08:03
NC-NL	23	41	12:17	10:15
NC-S	19	33	10:53	08:29
NJ	21	41	11:27	09:57
NM	29	51	08:02	06:40
OK	23	45	11:11	09:15
OR	30	32	09:09	08:09

Table A.6—Continued

	Push-Ups		1-Mile Run	
Site	Initial	Final	Initial	Final
PR	21	40	10:11	08:10
SC	29	*	10:00	*
TN	*	*	*	*
TX[a]	35	38	10:46	10:50
VA	14	40	10:17	08:35
WA	20	38	11:02	07:14
WI	20	41	09:18	07:56
WV	24	41	10:10	07:51
WY	28	44	09:31	07:50

NOTE: Information in this table was reported by the sites in July and August 2019 and covers Class 51.

* = did not report.

[a]Pull-ups; site does not collect data on push-ups.

Table A.7
Profile of Alaska Military Youth Academy

ALASKA MILITARY YOUTH ACADEMY, ESTABLISHED 1994

Graduates Since Inception: 5,800 **Program Type: Credit Recovery, High School Diploma, GED**

Staffing

	Instructional	Cadre	Admin.	Case Managers	Recruiters	Other
Number employed	7	29	9	5	3	13

Funding

	Federal Funding	State Funding	Other Funding
Classes 50 and 51	$3,885,630	$5,757,793	$0

Residential Performance

	Dates	Applied	Entered Pre-ChalleNGe	Graduated	Received GED/HiSET	Received HS Credits	Received HS Diploma
Class 50	Mar 2018–Jul 2018	248	227	187	26	137	1
Class 51	Sep 2018–Jan 2019	221	195	156	22	107	2

Table A.7—Continued

Physical Fitness

	Push-Ups		1-Mile Run		BMI	
	Initial	Final	Initial	Final	Initial	Final
Class 50	27	44	10:41	08:49	26.3	*
Class 51	25	42	10:10	09:01	25.6	*

Responsible Citizenship

	Voting		Selective Service	
	Eligible	Registered	Eligible	Registered
Class 50	38	37	25	24
Class 51	30	29	23	20

Service to Community

	Service Hours/Cadet	Dollar Value/ Hour	Total Value
Class 50	62	$27.88	$322,761
Class 51	49	$27.88	$212,754

Post-Residential Performance Status

	Graduated	Contacted	Placed	Education	Employment	Military	Multiple/ Other
Class 49							
Month 12	146	65	63	26	21	2	14
Class 50							
Month 1	187	77	76	69	3	0	5
Month 6	187	32	31	16	4	0	11
Month 12	187	N/A	N/A	N/A	N/A	N/A	N/A
Class 51							
Month 1	156	70	67	46	12	1	8
Month 6	156	N/A	N/A	N/A	N/A	N/A	N/A

NOTES: * = did not report. N/A = not applicable; follow-up period has not occurred.

Table A.8
Profile of Arkansas Youth ChalleNGe

ARKANSAS YOUTH CHALLENGE, ESTABLISHED 1993

Graduates Since Inception: 3,934 **Program Type: GED**

Staffing

	Instructional	Cadre	Admin.	Case Managers	Recruiters	Other
Number employed	5	23	19	4	2	2

Funding

	Federal Funding	State Funding	Other Funding
Classes 50 and 51	$2,512,500	$837,500	$0

Residential Performance

	Dates	Applied	Entered Pre-ChalleNGe	Graduated	Received GED/HiSET	Received HS Credits	Received HS Diploma
Class 50	Jan 2018–Jun 2018	216	135	80	21	0	0
Class 51	Jul 2018–Dec 2018	257	162	107	44	0	0

Physical Fitness

	Push-Ups		1-Mile Run		BMI	
	Initial	Final	Initial	Final	Initial	Final
Class 50	29	46	10:31	09:22	25.5	*
Class 51	31	47	10:26	10:05	24.4	*

Responsible Citizenship

	Voting		Selective Service	
	Eligible	Registered	Eligible	Registered
Class 50	21	20	40	39
Class 51	20	19	52	51

Service to Community

	Service Hours/Cadet	Dollar Value/ Hour	Total Value
Class 50	98	$20.49	$160,437
Class 51	71	$20.49	$154,945

Table A.8—Continued

Post-Residential Performance Status

	Graduated	Contacted	Placed	Education	Employment	Military	Multiple/ Other
Class 49							
Month 12	88	49	40	13	12	2	13
Class 50							
Month 1	80	66	32	16	12	0	10
Month 6	80	42	34	15	9	0	12
Month 12	80	36	28	9	6	0	16
Class 51							
Month 1	107	73	47	32	8	0	11
Month 6	107	51	42	17	9	0	11

NOTE: * = did not report.

Table A.9
Profile of Discovery ChalleNGe Academy (California)

DISCOVERY CHALLENGE ACADEMY, ESTABLISHED 2017

Graduates Since Inception: 480

Program Type: Credit Recovery, High School Diploma, GED

Staffing

	Instructional	Cadre	Admin.	Case Managers	Recruiters	Other
Number employed	6	25	8	3	1	4

Funding

	Federal Funding	State Funding	Other Funding
Classes 50 and 51	$8,560,000	$2,853,333	$3,846,157

Residential Performance

	Dates	Applied	Entered Pre-ChalleNGe	Graduated	Received GED/HiSET	Received HS Credits	Received HS Diploma
Class 50	Jan 2018–Jun 2018	200	133	120	0	86	34
Class 51	Jul 2018–Dec 2018	259	146	129	0	104	25

Table A.9—Continued

Physical Fitness

	Push-Ups		1-Mile Run		BMI	
	Initial	Final	Initial	Final	Initial	Final
Class 50	26	44	09:16	07:24	*	*
Class 51	29	47	09:35	07:41	*	*

Responsible Citizenship

	Voting		Selective Service	
	Eligible	Registered	Eligible	Registered
Class 50	28	28	28	28
Class 51	23	23	23	23

Service to Community

	Service Hours/Cadet	Dollar Value/Hr	Total Value
Class 50	44	$29.95	$158,136
Class 51	52	$29.95	$200,905

Post-Residential Performance Status

	Graduated	Contacted	Placed	Education	Employment	Military	Multiple/ Other
Class 49							
Month 12	126	126	116	66	25	10	15
Class 50							
Month 1	120	120	93	84	4	0	5
Month 6	120	120	100	73	13	1	13
Month 12	120	120	110	69	26	6	9
Class 51							
Month 1	129	129	107	86	11	0	14
Month 6	129	129	116	77	17	5	17

NOTE: * = did not report.

Table A.10
Profile of Sunburst Youth Academy (California)

SUNBURST YOUTH ACADEMY, ESTABLISHED 2008

Graduates Since Inception: 3,535	Program Type: Credit Recovery, High School Diploma, HiSET

Staffing

	Instructional	Cadre	Admin.	Case Managers	Recruiters	Other
Number employed	10	30	7	4	2	8

Funding

	Federal Funding	State Funding	Other Funding
Classes 50 and 51	$5,895,000	$1,965,000	$0

Residential Performance

	Dates	Applied	Entered Pre-ChalleNGe	Graduated	Received GED/HiSET	Received HS Credits	Received HS Diploma
Class 50	Jan 2018–Jun 2018	307	196	147	0	121	26
Class 51	Jul 2018–Dec 2018	420	209	186	0	161	25

Physical Fitness

	Push-Ups		1-Mile Run		BMI	
	Initial	Final	Initial	Final	Initial	Final
Class 50	26	47	10:12	08:04	26.5	26.1
Class 51	26	49	09:28	07:40	27.4	25.8

Responsible Citizenship

	Voting		Selective Service	
	Eligible	Registered	Eligible	Registered
Class 50	22	22	23	23
Class 51	30	30	34	34

Service to Community

	Service Hours/Cadet	Dollar Value/Hour	Total Value
Class 50	42	$29.95	$184,911
Class 51	43	$29.95	$239,540

Table A.10—Continued

Post-Residential Performance Status

	Graduated	Contacted	Placed	Education	Employment	Military	Multiple/ Other
Class 49							
Month 12	193	193	155	75	23	4	53
Class 50							
Month 1	147	138	138	99	25	4	10
Month 6	147	147	139	91	11	3	34
Month 12	147	112	112	62	30	8	12
Class 51							
Month 1	186	175	175	120	22	2	31
Month 6	186	N/A	N/A	N/A	N/A	N/A	N/A

NOTES: N/A = not applicable; follow-up period has not occurred.

Table A.11
Profile of Grizzly Youth Academy (California)

GRIZZLY YOUTH ACADEMY, ESTABLISHED 1998

Graduates Since Inception: 6,241	Program Type: Credit Recovery, High School Diploma, HiSET

Staffing

	Instructional	Cadre	Admin.	Case Managers	Recruiters	Other
Number employed	14	30	8	4	0	8

Funding

	Federal Funding	State Funding	Other Funding
Classes 50 and 51	$6,125,000	$2,041,666	$0

Residential Performance

	Dates	Applied	Entered pre-ChalleNGe	Graduated	Received GED/HiSET	Received HS Credits	Received HS Diploma
Class 50	Jan 2018–Jun 2018	311	235	184	8	123	53
Class 51	Jul 2018–Dec 2018	336	222	192	4	122	66

Table A.11—Continued

Physical Fitness

	Push-Ups		1-Mile Run		BMI	
	Initial	Final	Initial	Final	Initial	Final
Class 50	21	38	10:05	07:30	*	*
Class 51	20	36	09:23	07:41	*	*

Responsible Citizenship

	Voting		Selective Service	
	Eligible	Registered	Eligible	Registered
Class 50	42	31	34	24
Class 51	32	29	25	23

Service to Community

	Service Hours/Cadet	Dollar Value/Hour	Total Value
Class 50	48	$29.95	$264,518
Class 51	48	$29.95	$276,019

Post-Residential Performance Status

	Graduated	Contacted	Placed	Education	Employment	Military	Multiple/ Other
Class 49							
Month 12	195	187	159	40	59	5	55
Class 50							
Month 1	184	175	151	82	31	1	37
Month 6	184	171	140	58	33	2	47
Month 12	184	167	146	42	46	9	49
Class 51							
Month 1	193	191	166	118	15	0	33
Month 6	193	183	154	64	30	7	48

NOTE: * = did not report.

Table A.12
Profile of Capital Guardian Youth ChalleNGe Academy (District of Columbia)

CAPITAL GUARDIAN YOUTH CHALLENGE ACADEMY, ESTABLISHED 2007

Graduates Since Inception: 670 | **Program Type: GED**

Staffing

	Instructional	Cadre	Admin.	Case Managers	Recruiters	Other
Number Employed	5	23	7	6	2	15

Funding

	Federal Funding	State Funding	Other Funding
Classes 50 and 51	$2,256,121	$2,191,878	$0

Residential Performance

	Dates	Applied	Entered Pre-ChalleNGe	Graduated	Received GED/HiSET	Received HS Credits	Received HS Diploma
Class 50	Jan 2018–Jun 2018	158	50	31	14	0	0
Class 51	Jul 2018–Dec 2018	180	68	41	12	0	0

Physical Fitness

	Push-Ups		1-Mile Run		BMI	
	Initial	Final	Initial	Final	Initial	Final
Class 50	23	41	10:19	09:44	23.2	26.0
Class 51	26	39	12:05	11:53	24.0	25.9

Responsible Citizenship

	Voting		Selective Service	
	Eligible	Registered	Eligible	Registered
Class 50	9	9	9	9
Class 51	10	10	8	8

Service to Community

	Service Hours/Cadet	Dollar Value/Hour	Total Value
Class 50	64	$41.72	$82,190
Class 51	52	$41.72	$89,323

Table A.12—Continued

Post-Residential Performance Status

	Graduated	Contacted	Placed	Education	Employment	Military	Multiple/ Other
Class 49							
Month 12	45	45	18	8	6	2	2
Class 50							
Month 1	31	30	14	0	14	0	0
Month 6	31	23	16	6	7	0	3
Month 12	31	25	15	6	7	0	2
Class 51							
Month 1	41	38	10	8	1	0	1
Month 6	41	36	25	11	8	2	2

Table A.13
Profile of Florida Youth ChalleNGe Academy (Florida)

FLORIDA YOUTH CHALLENGE ACADEMY, ESTABLISHED 2001

Graduates Since Inception: 4,950

Program Type: Credit Recovery, GED, High School Diploma

Staffing

	Instructional	Cadre	Admin.	Case Managers	Recruiters	Other
Number employed	9	34	16	5	1	18

Funding

	Federal Funding	State Funding	Other Funding
Classes 50 and 51	$4,051,569	$1,350,523	$288,269

Residential Performance

	Dates	Applied	Entered Pre-ChalleNGe	Graduated	Received GED/HiSET	Received HS Credits	Received HS Diploma
Class 50	Jan 2018–Jun 2018	174	174	137	83	16	2
Class 51	Jul 2018–Dec 2018	214	207	159	116	24	8

Table A.13—Continued

Physical Fitness

	Push-Ups		1-Mile Run		BMI	
	Initial	Final	Initial	Final	Initial	Final
Class 50	18	34	10:24	08:22	25.0	25.2
Class 51	15	29	11:19	09:01	26.1	26.5

Responsible Citizenship

	Voting		Selective Service	
	Eligible	Registered	Eligible	Registered
Class 50	40	40	29	29
Class 51	38	38	35	35

Service to Community

	Service Hours/Cadet	Dollar Value/Hour	Total Value
Class 50	83	$24.04	$271,712
Class 51	79	$24.04	$300,055

Post-Residential Performance Status

	Graduated	Contacted	Placed	Education	Employment	Military	Multiple/ Other
Class 49							
Month 12	158	99	77	18	49	6	4
Class 50							
Month 1	137	96	62	16	38	0	8
Month 6	137	74	62	14	38	3	7
Month 12	137	88	77	18	47	6	6
Class 51							
Month 1	159	127	94	20	65	3	6
Month 6	159	130	110	18	84	2	2

Table A.14
Profile of Fort Gordon Youth Academy (Georgia)

FORT GORDON YOUTH CHALLENGE ACADEMY, ESTABLISHED 2000

Graduates Since Inception: 6,391	Program Type: Credit Recovery, High School Diploma, GED

Staffing

	Instructional	Cadre	Admin.	Case Managers	Recruiters	Other
Number employed	8	37	12	4	1	44

Funding

	Federal Funding	State Funding	Other Funding
Classes 50 and 51	$5,117,007	$1,705,669	$257,222

Residential Performance

	Dates	Applied	Entered Pre-ChalleNGe	Graduated	Received GED/ HiSET	Received HS Credits	Received HS Diploma
Class 50	Mar 2018–Aug 2018	306	252	182	57	0	36
Class 51	Sep 2018–Feb 2019	223	186	140	36	33	15

Physical Fitness

	Push-Ups		1-Mile Run		BMI	
	Initial	Final	Initial	Final	Initial	Final
Class 50	27	39	09:45	09:14	*	*
Class 51	32	36	10:15	09:34	*	*

Responsible Citizenship

	Voting		Selective Service	
	Eligible	Registered	Eligible	Registered
Class 50	51	0	81	81
Class 51	35	0	72	71

Service to Community

	Service Hours/Cadet	Dollar Value/Hour	Total Value
Class 50	56	$25.78	$260,573
Class 51	43	$25.78	$154,737

Table A.14—Continued

Post-Residential Performance Status

	Graduated	Contacted	Placed	Education	Employment	Military	Multiple/ Other
Class 49							
Month 12	135	80	44	10	31	2	1
Class 50							
Month 1	182	124	124	37	79	1	7
Month 6	182	155	155	36	104	9	6
Month 12	182	N/A	N/A	N/A	N/A	N/A	N/A
Class 51							
Month 1	140	94	94	23	63	3	5
Month 6	140	N/A	N/A	N/A	N/A	N/A	N/A

NOTES: * = did not report. N/A = not applicable; follow-up period has not occurred.

Table A.15
Profile of Fort Stewart Youth Academy (Georgia)

FORT STEWART YOUTH CHALLENGE ACADEMY, ESTABLISHED 1993

Graduates Since Inception: 10,137 | **Program Type: Credit Recovery, High School Diploma, GED**

Staffing

	Instructional	Cadre	Admin.	Case Managers	Recruiters	Other
Number employed	7	53	16	5	2	31

Funding

	Federal Funding	State Funding	Other Funding
Classes 50 and 51	$3,246,675	$1,916,142	$1,652,297

Residential Performance

	Dates	Applied	Entered Pre-ChalleNGe	Graduated	Received GED/HiSET	Received HS Credits	Received HS Diploma
Class 50	Jan 2018–Jun 2018	312	218	174	39	24	49
Class 51	Jul 2018–Dec 2018	335	260	199	20	28	62

Table A.15—Continued

Physical Fitness

	Push-Ups		1-Mile Run		BMI	
	Initial	Final	Initial	Final	Initial	Final
Class 50	31	45	10:46	09:06	24.7	25.8
Class 51	31	47	10:17	09:02	24.2	24.3

Responsible Citizenship

	Voting		Selective Service	
	Eligible	Registered	Eligible	Registered
Class 50	46	46	35	35
Class 51	42	42	36	36

Service to Community

	Service Hours/Cadet	Dollar Value/Hour	Total Value
Class 50	52	$25.78	$233,257
Class 51	73	$25.78	$372,624

Post-Residential Performance Status

	Graduated	Contacted	Placed	Education	Employment	Military	Multiple/ Other
Class 49							
Month 12	224	195	178	39	82	10	47
Class 50							
Month 1	174	162	136	86	43	3	4
Month 6	174	162	155	87	59	6	3
Month 12	174	155	145	56	80	6	3
Class 51							
Month 1	199	190	154	111	27	2	17
Month 6	199	177	155	77	44	5	15

Table A.16
Profile of Milledgeville Youth ChalleNGe Academy (Georgia)

MILLEDGEVILLE YOUTH CHALLENGE ACADEMY, ESTABLISHED 2016

Graduates Since Inception: 371	Program Type: Credit Recovery, High School Diploma, GED

Staffing

	Instructional	Cadre	Admin.	Case Managers	Recruiters	Other
Number employed	7	24	14	4	2	4

Funding

	Federal Funding	State Funding	Other Funding
Classes 50 and 51	$4,639,069	$1,546,356	$0

Residential Performance

	Dates	Applied	Entered Pre-ChalleNGe	Graduated	Received GED/HiSET	Received HS Credits	Received HS Diploma
Class 50	Nov 2017–Apr 2018	182	142	86	29	14	9
Class 51	May 2018–Oct 2018	189	189	102	30	28	18

Physical Fitness

	Push-Ups		1-Mile Run		BMI	
	Initial	Final	Initial	Final	Initial	Final
Class 50	43	49	09:41	08:13	*	*
Class 51	*	37	09:51	08:41	*	*

Responsible Citizenship

	Voting		Selective Service	
	Eligible	Registered	Eligible	Registered
Class 50	32	32	28	28
Class 51	32	32	21	21

Service to Community

	Service Hours/Cadet	Dollar Value/Hour	Total Value
Class 50	43	$25.78	$98,572
Class 51	44	$25.78	$116,174

Table A.16—Continued

Post-Residential Performance Status

	Graduated	Contacted	Placed	Education	Employment	Military	Multiple/ Other
Class 49							
Month 12	107	107	81	5	40	8	28
Class 50							
Month 1	86	78	49	13	29	3	4
Month 6	86	80	73	12	41	3	20
Month 12	86	82	75	12	39	7	19
Class 51							
Month 1	102	66	41	13	17	2	13
Month 6	102	96	82	22	43	7	6

NOTE: * = did not report.

Table A.17
Profile of ChalleNGe Academy at Barbers Point (Hawaii)

HAWAII YOUTH CHALLENGE ACADEMY AT BARBERS POINT, ESTABLISHED 1993

Graduates Since Inception: 4,449 **Program Type: High School Diploma, HiSET**

Staffing

	Instructional	Cadre	Admin.	Case Managers	Recruiters	Other
Number employed	9	20	11	1	1	5

Funding

	Federal Funding	State Funding	Other Funding
Classes 50 and 51	$3,187,500	$1,062,500	$0

Residential Performance

	Dates	Applied	Entered Pre-ChalleNGe	Graduated	Received GED/HiSET	Received HS Credits	Received HS Diploma
Class 50	Jan 2018–Jun 2018	130	105	94	0	0	94
Class 51	Jul 2018–Dec 2018	122	101	79	15	0	64

Table A.17—Continued

Physical Fitness

	Push-Ups		1-Mile Run		BMI	
	Initial	Final	Initial	Final	Initial	Final
Class 50	27	47	10:29	07:52	25.1	*
Class 51	32	53	10:51	08:09	25.2	*

Responsible Citizenship

	Voting		Selective Service	
	Eligible	Registered	Eligible	Registered
Class 50	94	94	72	72
Class 51	79	79	56	56

Service to Community

	Service Hours/Cadet	Dollar Value/ Hour	Total Value
Class 50	156	$26.87	$393,969
Class 51	121	$26.87	$256,394

Post-Residential Performance Status

	Graduated	Contacted	Placed	Education	Employment	Military	Multiple/ Other
Class 49							
Month 12	105	87	53	1	28	8	16
Class 50							
Month 1	94	94	59	12	39	1	8
Month 6	94	84	48	2	39	0	8
Month 12	94	66	44	4	34	4	2
Class 51							
Month 1	79	65	34	2	29	1	8
Month 6	79	35	22	0	21	0	0

NOTE: * = did not report.

Table A.18
Profile of Youth Academy at Hilo (Hawaii)

HAWAII YOUTH CHALLENGE ACADEMY AT HILO, ESTABLISHED 2011

Graduates Since Inception: 887 | **Program Type: Credit Recovery, HiSET**

Staffing

	Instructional	Cadre	Admin.	Case Managers	Recruiters	Other
Number employed	3	15	11	1	3	0

Funding

	Federal Funding	State Funding	Other Funding
Classes 50 and 51	$1,912,500	$637,500	$0

Residential Performance

	Dates	Applied	Entered Pre-ChalleNGe	Graduated	Received GED/ HiSET	Received HS Credits	Received HS Diploma
Class 50	Jan 2018–Jun 2018	85	74	68	0	0	64
Class 51	Jul 2018–Dec 2018	127	84	63	0	0	61

Physical Fitness

	Push-Ups		1-Mile Run		BMI	
	Initial	Final	Initial	Final	Initial	Final
Class 50	33	41	09:29	09:09	24.2	23.7
Class 51	34	48	08:51	09:05	25.1	24.2

Responsible Citizenship

	Voting		Selective Service	
	Eligible	Registered	Eligible	Registered
Class 50	62	62	42	42
Class 51	59	59	41	41

Service to Community

	Service Hours/Cadet	Dollar Value/ Hour	Total Value
Class 50	117	$26.87	$213,778
Class 51	116	$26.87	$199,483

Table A.18—Continued

Post-Residential Performance Status

	Graduated	Contacted	Placed	Education	Employment	Military	Multiple/ Other
Class 49							
Month 12	67	65	25	1	13	8	3
Class 50							
Month 1	68	41	16	3	10	2	1
Month 6	68	65	19	2	15	1	1
Month 12	68	61	29	3	20	4	2
Class 51							
Month 1	63	60	13	0	9	3	1
Month 6	63	57	18	1	12	5	0

Table A.19
Profile of Idaho Youth ChalleNGe Academy

IDAHO YOUTH CHALLENGE ACADEMY, ESTABLISHED 2014

Graduates Since Inception: 1,003	Program Type: Credit Recovery, High School Diploma, GED

Staffing

	Instructional	Cadre	Admin.	Case Managers	Recruiters	Other
Number employed	6	25	8	4	2	13

Funding

	Federal Funding	State Funding	Other Funding
Classes 50 and 51	$2,955,736	$984,072	$556,688

Residential Performance

	Dates	Applied	Entered Pre-ChalleNGe	Graduated	Received GED/ HiSET	Received HS Credits	Received HS Diploma
Class 50	Jan 2018–Jun 2018	166	129	111	0	97	14
Class 51	Jul 2018–Dec 2018	170	150	129	0	112	17

Table A.19—Continued

Physical Fitness

	Push-Ups		1-Mile Run		BMI	
	Initial	Final	Initial	Final	Initial	Final
Class 50	22	44	09:35	07:37	25.3	24.6
Class 51	27	40	10:08	07:43	25.0	24.4

Responsible Citizenship

	Voting		Selective Service	
	Eligible	Registered	Eligible	Registered
Class 50	13	13	32	32
Class 51	26	26	37	37

Service to Community

	Service Hours/Cadet	Dollar Value/Hour	Total Value
Class 50	48	$22.14	$116,979
Class 51	48	$22.14	$136,805

Post-Residential Performance Status

	Graduated	Contacted	Placed	Education	Employment	Military	Multiple/ Other
Class 49							
Month 12	115	99	83	30	16	3	34
Class 50							
Month 1	111	99	23	8	9	1	22
Month 6	111	107	80	34	21	4	34
Month 12	111	104	80	22	28	5	39
Class 51							
Month 1	129	114	41	33	2	2	22
Month 6	129	96	55	16	11	5	18

Table A.20
Profile of Lincoln's ChalleNGe Academy (Illinois)

LINCOLN'S CHALLENGE ACADEMY, ESTABLISHED 1993	
Graduates Since Inception: 15,401	Program Type: GED, Credit Recovery, High School Diploma

Staffing

	Instructional	Cadre	Admin.	Case Managers	Recruiters	Other
Number employed	6	36	27	4	6	5

Funding

	Federal Funding	State Funding	Other Funding
Classes 50 and 51	$4,825,000	$1,382,501	$0

Residential Performance

	Dates	Applied	Entered Pre-ChalleNGe	Graduated	Received GED/ HiSET	Received HS Credits	Received HS Diploma
Class 50	Feb 2018–Jul 2018	284	236	116	65	0	0
Class 51	Jul 2018–Dec 2018	272	205	123	67	0	0

Physical Fitness

	Push-Ups		1-Mile Run		BMI	
	Initial	Final	Initial	Final	Initial	Final
Class 50	21	41	11:20	08:56	*	*
Class 51	27	43	11:53	09:09	*	*

Responsible Citizenship

	Voting		Selective Service	
	Eligible	Registered	Eligible	Registered
Class 50	19	19	19	19
Class 51	21	21	21	21

Service to Community

	Service Hours/Cadet	Dollar Value/Hour	Total Value
Class 50	51	$26.89	$159,050
Class 51	74	$26.89	$246,609

Table A.20—Continued

Post-Residential Performance Status

	Graduated	Contacted	Placed	Education	Employment	Military	Multiple/ Other
Class 49							
Month 12	138	134	21	1	10	2	8
Class 50							
Month 1	116	47	26	2	16	2	11
Month 6	116	47	33	4	23	2	8
Month 12	116	27	16	2	9	3	4
Class 51							
Month 1	123	78	40	10	17	3	19
Month 6	123	62	42	8	19	4	10

NOTE: * = did not report.

Table A.21
Profile of Hoosier Youth ChalleNGe Academy (Indiana)

HOOSIER YOUTH CHALLENGE ACADEMY, ESTABLISHED 2007

Graduates Since Inception: 1,771 **Program Type: TASC**

Staffing

	Instructional	Cadre	Admin.	Case Managers	Recruiters	Other
Number employed	5	21	7	7	4	2

Funding

	Federal Funding	State Funding	Other Funding
Classes 50 and 51	$3,458,345	$1,152,781	$0

Residential Performance

	Dates	Applied	Entered Pre-ChalleNGe	Graduated	Received GED/HiSET/ TASC	Received HS Credits	Received HS Diploma
Class 50	Jan 2018–Jun 2018	145	122	72	53	0	0
Class 51	Jul 2018–Dec 2018	129	93	61	42	0	0

Table A.21—Continued

Physical Fitness

	Push-Ups		1-Mile Run		BMI	
	Initial	Final	Initial	Final	Initial	Final
Class 50	22	42	10:27	08:15	*	*
Class 51	32	46	12:13	09:49	*	*

Responsible Citizenship

	Voting		Selective Service	
	Eligible	Registered	Eligible	Registered
Class 50	5	0	29	29
Class 51	3	0	24	24

Service to Community

	Service Hours/Cadet	Dollar Value/Hour	Total Value
Class 50	56	$24.13	$96,771
Class 51	52	$24.13	$75,952

Post-Residential Performance Status

	Graduated	Contacted	Placed	Education	Employment	Military	Multiple/ Other
Class 49							
Month 12	97	97	28	10	15	1	2
Class 50							
Month 1	72	63	7	1	5	0	1
Month 6	72	68	21	6	12	0	3
Month 12	72	59	10	4	5	0	1
Class 51							
Month 1	61	45	3	2	1	0	0
Month 6	61	43	6	2	4	0	0

NOTE: * = did not report.

Table A.22
Profile of Bluegrass ChalleNGe Academy (Kentucky)

BLUEGRASS CHALLENGE ACADEMY, ESTABLISHED 1999

Graduates Since Inception: 3,219	Program Type: Credit Recovery, GED, High School Diploma

Staffing

	Instructional	Cadre	Admin.	Case Managers	Recruiters	Other
Number employed	5	27	9	1	4	8

Funding

	Federal Funding	State Funding	Other Funding
Classes 50 and 51	$2,673,750	$891,250	$0

Residential Performance

	Dates	Applied	Entered Pre-ChalleNGe	Graduated	Received GED/ HiSET	Received HS Credits	Received HS Diploma
Class 50	Apr 2018–Sep 2018	140	116	65	0	49	12
Class 51	Oct 2018–Mar 2019	83	65	41	0	35	6

Physical Fitness

	Push-Ups		1-Mile Run		BMI	
	Initial	Final	Initial	Final	Initial	Final
Class 50	29	38	12:32	10:39	24.9	23.8
Class 51	23	47	10:14	09:06	24.0	23.3

Responsible Citizenship

	Voting		Selective Service	
	Eligible	Registered	Eligible	Registered
Class 50	10	10	10	10
Class 51	16	16	16	16

Service to Community

	Service Hours/Cadet	Dollar Value/Hour	Total Value
Class 50	65	$21.42	$89,135
Class 51	75	$21.42	$65,849

Table A.22—Continued

Post-Residential Performance Status

	Graduated	Contacted	Placed	Education	Employment	Military	Multiple/ Other
Class 49							
Month 12	72	73	64	51	11	1	1
Class 50							
Month 1	65	64	64	55	8	0	1
Month 6	65	63	58	48	6	1	3
Month 12	65	N/A	N/A	N/A	N/A	N/A	N/A
Class 51							
Month 1	41	41	41	41	0	0	0
Month 6	41	39	33	23	7	0	3

NOTES: N/A = not applicable; follow-up period has not occurred.

Table A.23
Profile of Appalachian ChalleNGe Program (Kentucky)

APPALACHIAN CHALLENGE PROGRAM, ESTABLISHED 2012

Graduates Since Inception: 1,048 | **Program Type: Credit Recovery, GED**

Staffing

	Instructional	Cadre	Admin.	Case Managers	Recruiters	Other
Number employed	5	27	11	2	3	5

Funding

	Federal Funding	State Funding	Other Funding
Classes 50 and 51	$3,243,771	$1,081,257	$0

Residential Performance

	Dates	Applied	Entered Pre-ChalleNGe	Graduated	Received GED/ HiSET	Received HS Credits	Received HS Diploma
Class 50	Jan 2018–Jun 2018	149	119	74	0	73	0
Class 51	Jul 2018–Dec 2018	111	90	90	0	89	0

Table A.23—Continued

Physical Fitness

	Push-Ups		1-Mile Run		BMI	
	Initial	Final	Initial	Final	Initial	Final
Class 50	30	52	09:34	08:55	26.6	26.6
Class 51	23	47	09:51	08:34	24.9	24.8

Responsible Citizenship

	Voting		Selective Service	
	Eligible	Registered	Eligible	Registered
Class 50	14	14	15	15
Class 51	7	7	7	7

Service to Community

	Service Hours/Cadet	Dollar Value/ Hour	Total Value
Class 50	57	$24.19	$101,407
Class 51	44	$24.19	$71,717

Post-Residential Performance Status

	Graduated	Contacted	Placed	Education	Employment	Military	Multiple/ Other
Class 49							
Month 12	82	68	68	59	7	2	0
Class 50							
Month 1	74	74	59	49	5	2	3
Month 6	74	74	58	40	2	2	14
Month 12	74	74	55	45	2	2	6
Class 51							
Month 1	90	67	57	45	2	3	7
Month 6	90	67	57	44	2	3	8

Table A.24
Profile of Camp Beauregard (Louisiana)

LOUISIANA YOUTH CHALLENGE PROGRAM—CAMP BEAUREGARD, ESTABLISHED 1993	
Graduates Since Inception: 10,517	Program Type: HiSET, Credit Recovery

Staffing

	Instructional	Cadre	Admin.	Case Managers	Recruiters	Other
Number employed	15	46	19	12	2	28

Funding

	Federal Funding	State Funding	Other Funding
Classes 50 and 51	$6,937,500	$2,312,500	$0

Residential Performance

	Dates	Applied	Entered Pre-ChalleNGe	Graduated	Received GED/HiSET	Received HS Credits	Received HS Diploma
Class 50	Jan 2018–Jun 2018	424	276	204	65	0	0
Class 51	Jul 2018–Dec 2018	427	296	201	68	0	0

Physical Fitness

	Push-Ups		1-Mile Run		BMI	
	Initial	Final	Initial	Final	Initial	Final
Class 50	25	44	09:34	08:08	24.6	*
Class 51	32	36	09:56	09:27	25.5	*

Responsible Citizenship

	Voting		Selective Service	
	Eligible	Registered	Eligible	Registered
Class 50	32	32	94	94
Class 51	28	28	93	93

Service to Community

	Service Hours/Cadet	Dollar Value/Hour	Total Value
Class 50	50	$22.76	$230,377
Class 51	52	$22.76	$238,183

Table A.24—Continued

Post-Residential Performance Status

	Graduated	Contacted	Placed	Education	Employment	Military	Multiple/ Other
Class 49							
Month 12	250	212	177	28	88	9	52
Class 50							
Month 1	204	204	174	27	116	0	31
Month 6	204	197	172	44	86	0	43
Month 12	204	173	152	26	85	5	37
Class 51							
Month 1	201	196	170	43	84	3	40
Month 6	201	189	160	38	65	8	45

NOTE: * = did not report.

Table A.25
Profile of Camp Minden (Louisiana)

LOUISIANA YOUTH CHALLENGE PROGRAM—CAMP MINDEN, ESTABLISHED 2002

Graduates Since Inception: 5,459 | Program Type: HiSET, Credit Recovery

Staffing

	Instructional	Cadre	Admin.	Case Managers	Recruiters	Other
Number employed	13	48	18	8	0	23

Funding

	Federal Funding	State Funding	Other Funding
Classes 50 and 51	$5,100,000	$1,700,000	$0

Residential Performance

	Dates	Applied	Entered Pre-ChalleNGe	Graduated	Received GED/ HiSET	Received HS Credits	Received HS Diploma
Class 50	Feb 2018–Jul 2018	332	254	204	58	7	1
Class 51	Aug 2018–Jan 2019	319	233	174	53	0	0

Table A.25—Continued

Physical Fitness

	Push-Ups		1-Mile Run		BMI	
	Initial	Final	Initial	Final	Initial	Final
Class 50	18	34	10:42	07:53	25.9	26.0
Class 51	21	35	12:30	08:31	25.8	25.1

Responsible Citizenship

	Voting		Selective Service	
	Eligible	Registered	Eligible	Registered
Class 50	21	21	57	57
Class 51	22	22	51	50

Service to Community

	Service Hours/Cadet	Dollar Value/ Hour	Total Value
Class 50	50	$22.76	$235,016
Class 51	66	$22.76	$263,276

Post-Residential Performance Status

	Graduated	Contacted	Placed	Education	Employment	Military	Multiple/ Other
Class 49							
Month 12	213	212	191	49	75	9	58
Class 50							
Month 1	204	204	191	68	46	4	73
Month 6	204	204	183	66	66	6	45
Month 12	204	204	172	49	68	11	44
Class 51							
Month 1	174	174	161	52	43	5	61
Month 6	174	174	161	49	59	5	10

Table A.26
Profile of Gillis Long (Louisiana)

LOUISIANA YOUTH CHALLENGE PROGRAM—GILLIS LONG, ESTABLISHED 1999

Graduates Since Inception: 8,601 | **Program Type: HiSET, Credit Recovery**

Staffing

	Instructional	Cadre	Admin.	Case Managers	Recruiters	Other
Number employed	15	45	19	12	1	33

Funding

	Federal Funding	State Funding	Other Funding
Classes 50 and 51	$6,937,500	$2,312,500	$461,688

Residential Performance

	Dates	Applied	Entered Pre-ChalleNGe	Graduated	Received GED/HiSET	Received HS Credits	Received HS Diploma
Class 50	Apr 2018–Sep 2018	459	333	229	82	0	0
Class 51	Oct 2018–Mar 2019	379	276	188	71	0	0

Physical Fitness

	Push-Ups		1-Mile Run		BMI	
	Initial	Final	Initial	Final	Initial	Final
Class 50	29	43	10:16	09:32	*	*
Class 51	27	41	10:51	09:23	*	*

Responsible Citizenship

	Voting		Selective Service	
	Eligible	Registered	Eligible	Registered
Class 50	33	33	26	26
Class 51	29	25	23	19

Service to Community

	Service Hours/Cadet	Dollar Value/ Hour	Total Value
Class 50	46	$22.76	$239,367
Class 51	69	$22.76	$294,753

Table A.26—Continued

Post-Residential Performance Status

	Graduated	Contacted	Placed	Education	Employment	Military	Multiple/ Other
Class 49							
Month 12	214	176	170	24	60	11	75
Class 50							
Month 1	229	204	195	39	46	9	103
Month 6	229	208	193	51	52	11	81
Month 12	229	N/A	N/A	N/A	N/A	N/A	N/A
Class 51							
Month 1	188	150	143	36	59	2	46
Month 6	188	N/A	N/A	N/A	N/A	N/A	N/A

NOTES: * = did not report. N/A = not applicable; follow-up period has not occurred.

Table A.27
Profile of Freestate ChalleNGe Academy (Maryland)

FREESTATE CHALLENGE ACADEMY, ESTABLISHED 1993

Graduates Since Inception: 4,453 | **Program Type: High School Diploma**

Staffing

	Instructional	Cadre	Admin.	Case Managers	Recruiters	Other
Number employed	4	28	4	8	2	5

Funding

	Federal Funding	State Funding	Other Funding
Classes 50 and 51	$3,173,578	$1,057,859	$0

Residential Performance

	Dates	Applied	Entered Pre-ChalleNGe	Graduated	Received GED/HiSET	Received HS Credits	Received HS Diploma
Class 50	Jan 2018–Jun 2018	206	148	71	0	0	40
Class 51	Jul 2018–Dec 2018	273	163	98	0	0	65

Table A.27—Continued

Physical Fitness

	Push-Ups		1-Mile Run		BMI	
	Initial	Final	Initial	Final	Initial	Final
Class 50	26	40	11:39	10:11	27.8	*
Class 51	19	52	09:38	07:57	24.3	*

Responsible Citizenship

	Voting		Selective Service	
	Eligible	Registered	Eligible	Registered
Class 50	24	24	71	71
Class 51	17	17	98	98

Service to Community

	Service Hours/Cadet	Dollar Value/Hour	Total Value
Class 50	54	$28.65	$110,174
Class 51	46	$28.65	$129,627

Post-Residential Performance Status

	Graduated	Contacted	Placed	Education	Employment	Military	Multiple/ Other
Class 49							
Month 12	102	99	71	15	53	1	2
Class 50							
Month 1	71	71	34	3	31	0	0
Month 6	71	71	42	9	32	1	0
Month 12	71	71	44	9	34	1	0
Class 51							
Month 1	98	98	30	3	22	4	1
Month 6	98	98	68	9	54	5	0

NOTE: * = did not report.

Table A.28
Profile of Michigan Youth ChalleNGe Academy

MICHIGAN YOUTH CHALLENGE ACADEMY, ESTABLISHED 1999

Graduates Since Inception: 3,820 | **Program Type: Credit Recovery, High School Diploma, GED**

Staffing

	Instructional	Cadre	Admin.	Case Managers	Recruiters	Other
Number employed	8	31	11	4	2	2

Funding

	Federal Funding	State Funding	Other Funding
Classes 50 and 51	$3,385,500	$1,128,500	$0

Residential Performance

	Dates	Applied	Entered Pre-ChalleNGe	Graduated	Received GED/ HiSET	Received HS Credits	Received HS Diploma
Class 50	Jan 2018–Jun 2018	196	143	115	0	89	26
Class 51	Jul 2018–Dec 2018	251	154	118	0	75	43

Physical Fitness

	Push-Ups		1-Mile Run		BMI	
	Initial	Final	Initial	Final	Initial	Final
Class 50	30	47	09:20	07:55	24.9	*
Class 51	28	47	08:50	07:57	*	*

Responsible Citizenship

	Voting		Selective Service	
	Eligible	Registered	Eligible	Registered
Class 50	22	22	32	32
Class 51	20	20	25	25

Service to Community

	Service Hours/Cadet	Dollar Value/Hr	Total Value
Class 50	57	$24.85	$161,637
Class 51	64	$24.85	$188,279

Table A.28—Continued

Post-Residential Performance Status

	Graduated	Contacted	Placed	Education	Employment	Military	Multiple/ Other
Class 49							
Month 12	119	37	32	8	17	1	6
Class 50							
Month 1	115	69	52	17	18	3	15
Month 6	115	34	30	9	9	3	9
Month 12	115	39	32	0	24	2	7
Class 51							
Month 1	118	75	60	42	14	1	9
Month 6	118	18	15	0	9	1	4

NOTE: * = did not report.

Table A.29
Profile of Mississippi Youth ChalleNGe Academy

MISSISSIPPI YOUTH CHALLENGE ACADEMY, ESTABLISHED 1994

Graduates Since Inception: 9,591 — Program Type: High School Diploma

Staffing

	Instructional	Cadre	Admin.	Case Managers	Recruiters	Other
Number employed	10	47	19	7	5	26

Funding

	Federal Funding	State Funding	Other Funding
Classes 50 and 51	$4,350,000	$1,450,000	$0

Residential Performance

	Dates	Applied	Entered Pre-ChalleNGe	Graduated	Received GED/HiSET	Received HS Credits	Received HS Diploma
Class 50	Jan 2018–Jun 2018	456	291	192	0	0	119
Class 51	Jul 2018–Dec 2018	495	286	201	0	0	123

Table A.29—Continued

Physical Fitness

	Push-Ups		1-Mile Run		BMI	
	Initial	Final	Initial	Final	Initial	Final
Class 50	19	45	11:42	07:56	25.1	*
Class 51	24	50	11:05	07:51	25.0	*

Responsible Citizenship

	Voting		Selective Service	
	Eligible	Registered	Eligible	Registered
Class 50	39	39	54	54
Class 51	40	40	79	79

Service to Community

	Service Hours/Cadet	Dollar Value/Hour	Total Value
Class 50	65	$19.70	$245,856
Class 51	72	$19.70	$285,098

Post-Residential Performance Status

	Graduated	Contacted	Placed	Education	Employment	Military	Multiple/ Other
Class 49							
Month 12	200	178	172	36	84	14	38
Class 50							
Month 1	192	175	137	29	76	4	30
Month 6	192	177	167	32	80	16	42
Month 12	192	166	158	24	82	18	37
Class 51							
Month 1	201	196	137	43	59	7	39
Month 6	201	185	173	32	91	13	24

NOTE: * = did not report.

Table A.30
Profile of Montana Youth ChalleNGe Academy

MONTANA YOUTH CHALLENGE ACADEMY, ESTABLISHED 1999

Graduates Since Inception: 3,009 | **Program Type: Credit Recovery, HiSET**

Staffing

	Instructional	Cadre	Admin.	Case Managers	Recruiters	Other
Number employed	6	29	8	6	4	5

Funding

	Federal Funding	State Funding	Other Funding
Classes 50 and 51	$3,628,100	$1,209,367	$208,409

Residential Performance

	Dates	Applied	Entered Pre-ChalleNGe	Graduated	Received GED/ HiSET	Received HS Credits	Received HS Diploma
Class 50	Jan 2018–Jun 2018	182	147	113	57	0	0
Class 51	Jul 2018–Dec 2018	153	133	102	52	0	0

Physical Fitness

	Push-Ups		1-Mile Run		BMI	
	Initial	Final	Initial	Final	Initial	Final
Class 50	21	39	11:04	08:12	24.2	*
Class 51	22	39	10:51	08:03	24.1	*

Responsible Citizenship

	Voting		Selective Service	
	Eligible	Registered	Eligible	Registered
Class 50	17	17	29	29
Class 51	23	23	28	28

Service to Community

	Service Hours/Cadet	Dollar Value/ Hour	Total Value
Class 50	58	$23.09	$150,836
Class 51	62	$23.09	$146,563

Table A.30—Continued

Post-Residential Performance Status

	Graduated	Contacted	Placed	Education	Employment	Military	Multiple/ Other
Class 49							
Month 12	80	68	66	10	19	1	36
Class 50							
Month 1	113	109	87	31	40	4	18
Month 6	113	110	90	32	42	6	13
Month 12	113	106	88	28	45	5	15
Class 51							
Month 1	102	100	84	48	20	6	10
Month 6	102	101	87	29	32	9	13

NOTE: * = did not report.

Table A.31
Profile of Tarheel ChalleNGe Academy–New London (North Carolina)

TARHEEL CHALLENGE ACADEMY—NEW LONDON, ESTABLISHED 2015	
Graduates Since Inception: 582	Program Type: High School Diploma, Credit Recovery, HiSET, GED

Staffing

	Instructional	Cadre	Admin.	Case Managers	Recruiters	Other
Number employed	6	21	13	3	2	10

Funding

	Federal Funding	State Funding	Other Funding
Classes 50 and 51	$3,002,291	$1,002,299	$143,796

Residential Performance

	Dates	Applied	Entered Pre-ChalleNGe	Graduated	Received GED/HiSET	Received HS Credits	Received HS Diploma
Class 50	Apr 2018–Sep 2018	340	156	113	44	0	24
Class 51	Oct 2018–Apr 2019	344	148	109	53	0	26

Table A.31—Continued

Physical Fitness

	Push-Ups		1-Mile Run		BMI	
	Initial	Final	Initial	Final	Initial	Final
Class 50	21	36	12:38	10:27	25.2	24.3
Class 51	23	41	12:17	10:15	24.4	23.9

Responsible Citizenship

	Voting		Selective Service	
	Eligible	Registered	Eligible	Registered
Class 50	44	44	18	18
Class 51	21	21	38	38

Service to Community

	Service Hours/Cadet	Dollar Value/Hr	Total Value
Class 50	87	$24.19	$236,358
Class 51	91	$24.19	$239,545

Post-Residential Performance Status

	Graduated	Contacted	Placed	Education	Employment	Military	Multiple/ Other
Class 49							
Month 12	99	99	38	10	18	1	9
Class 50							
Month 1	113	104	25	5	13	0	7
Month 6	113	109	17	1	11	3	2
Month 12	113	N/A	N/A	N/A	N/A	N/A	N/A
Class 51							
Month 1	109	104	26	2	19	1	4
Month 6	109	N/A	N/A	N/A	N/A	N/A	N/A

NOTES: N/A = not applicable; follow-up period has not occurred.

Table A.32
Profile of Tarheel ChalleNGe Academy–Salemburg (North Carolina)

TARHEEL CHALLENGE ACADEMY—SALEMBURG, ESTABLISHED 1994	
Graduates Since Inception: 4,926	Program Type: High School Diploma, Credit Recovery, HiSET, GED

Staffing

	Instructional	Cadre	Admin.	Case Managers	Recruiters	Other
Number employed	10	32	22	3	2	1

Funding

	Federal Funding	State Funding	Other Funding
Classes 50 and 51	$2,625,000	$875,000	$0

Residential Performance

	Dates	Applied	Entered Pre-ChalleNGe	Graduated	Received GED/ HiSET	Received HS Credits	Received HS Diploma
Class 50	Jan 2018–Jun 2018	388	163	103	49	0	27
Class 51	Jul 2018–Dec 2018	333	151	110	62	0	11

Physical Fitness

	Push-Ups		1-Mile Run		BMI	
	Initial	Final	Initial	Final	Initial	Final
Class 50	21	36	11:29	08:14	25.7	25.0
Class 51	19	33	10:53	08:29	24.7	24.4

Responsible Citizenship

	Voting		Selective Service	
	Eligible	Registered	Eligible	Registered
Class 50	87	87	15	15
Class 51	24	24	19	19

Service to Community

	Service Hours/Cadet	Dollar Value/Hour	Total Value
Class 50	81	$24.19	$201,169
Class 51	67	$24.19	$178,174

Table A.32—Continued

Post-Residential Performance Status

	Graduated	Contacted	Placed	Education	Employment	Military	Multiple/ Other
Class 49							
Month 12	105	90	77	13	40	2	22
Class 50							
Month 1	103	103	70	4	44	0	37
Month 6	103	103	91	24	54	3	10
Month 12	103	103	51	6	23	9	28
Class 51							
Month 1	110	110	87	7	42	0	38
Month 6	110	110	69	15	32	7	0

Table A.33
Profile of New Jersey Youth ChalleNGe Academy

NEW JERSEY YOUTH CHALLENGE ACADEMY, ESTABLISHED 1994

Graduates Since Inception: 4,085 **Program Type: GED**

Staffing

	Instructional	Cadre	Admin.	Case Managers	Recruiters	Other
Number employed	4	22	7	2	3	2

Funding

	Federal Funding	State Funding	Other Funding
Classes 50 and 51	$3,000,000	$1,000,000	$0

Residential Performance

	Dates	Applied	Entered Pre-ChalleNGe	Graduated	Received GED/HiSET	Received HS Credits	Received HS Diploma
Class 50	Apr 2018–Sep 2018	283	128	74	4	0	0
Class 51	Oct 2018–Mar 2019	269	126	80	19	0	0

Table A.33—Continued

Physical Fitness

	Push-Ups		1-Mile Run		BMI	
	Initial	Final	Initial	Final	Initial	Final
Class 50	29	39	11:38	09:46	*	*
Class 51	21	41	11:27	09:57	*	*

Responsible Citizenship

	Voting		Selective Service	
	Eligible	Registered	Eligible	Registered
Class 50	11	11	6	6
Class 51	16	16	14	14

Service to Community

	Service Hours/Cadet	Dollar Value/Hr	Total Value
Class 50	48	$28.82	$102,369
Class 51	42	$28.82	$96,835

Post-Residential Performance Status

	Graduated	Contacted	Placed	Education	Employment	Military	Multiple/ Other
Class 49							
Month 12	60	49	48	10	17	4	17
Class 50							
Month 1	74	19	9	4	5	0	0
Month 6	74	62	43	21	19	2	3
Month 12	74	N/A	N/A	N/A	N/A	N/A	N/A
Class 51							
Month 1	80	53	17	5	10	1	2
Month 6	80	N/A	N/A	N/A	N/A	N/A	N/A

NOTES: * = did not report. N/A = not applicable; follow-up period has not occurred.

Table A.34
Profile of New Mexico Youth ChalleNGe Academy

NEW MEXICO YOUTH CHALLENGE ACADEMY, ESTABLISHED 2001

Graduates Since Inception: 2,742 | **Program Type: HiSET**

Staffing

	Instructional	Cadre	Admin.	Case Managers	Recruiters	Other
Number employed	5	17	11	3	3	5

Funding

	Federal Funding	State Funding	Other Funding
Classes 50 and 51	$2,901,563	$892,188	$142,600

Residential Performance

	Dates	Applied	Entered Pre-ChalleNGe	Graduated	Received GED/ HiSET	Received HS Credits	Received HS Diploma
Class 50	Jan 2018–Jun 2018	174	142	112	66	0	0
Class 51	Jul 2018–Dec 2018	139	110	85	53	0	0

Physical Fitness

	Push-Ups		1-Mile Run		BMI	
	Initial	Final	Initial	Final	Initial	Final
Class 50	28	47	08:12	06:38	23.6	*
Class 51	29	51	08:02	06:40	*	*

Responsible Citizenship

	Voting		Selective Service	
	Eligible	Registered	Eligible	Registered
Class 50	20	20	52	52
Class 51	16	16	40	40

Service to Community

	Service Hours/Cadet	Dollar Value/ Hour	Total Value
Class 50	61	$21.20	$144,838
Class 51	54	$21.20	$97,308

Table A.34—Continued

Post-Residential Performance Status

	Graduated	Contacted	Placed	Education	Employment	Military	Multiple/ Other
Class 49							
Month 12	107	88	51	2	15	2	32
Class 50							
Month 1	112	90	66	1	46	1	29
Month 6	112	72	27	2	18	0	22
Month 12	112	10	4	0	4	0	1
Class 51							
Month 1	85	78	37	1	21	5	25
Month 6	85	16	10	2	6	2	0

NOTE: * = did not report.

Table A.35
Profile of Thunderbird Youth Academy (Oklahoma)

THUNDERBIRD YOUTH ACADEMY, ESTABLISHED 1993

Graduates Since Inception: 4,845 | Program Type: Credit Recovery, GED, High School Diploma

Staffing

	Instructional	Cadre	Admin.	Case Managers	Recruiters	Other
Number employed	6	24	13	4	4	14

Funding

	Federal Funding	State Funding	Other Funding
Classes 50 and 51	$3,214,418	$1,041,397	$49,000

Residential Performance

	Dates	Applied	Entered Pre-ChalleNGe	Graduated	Received GED/HiSET	Received HS Credits	Received HS Diploma
Class 50	Jan 2018–Jun 2018	455	164	103	11	81	11
Class 51	Jul 2018–Dec 2018	396	181	105	5	92	8

Table A.35—Continued

Physical Fitness

	Push-Ups		1-Mile Run		BMI	
	Initial	Final	Initial	Final	Initial	Final
Class 50	26	43	11:16	10:03	25.1	25.0
Class 51	23	45	11:11	09:15	26.2	25.5

Responsible Citizenship

	Voting		Selective Service	
	Eligible	Registered	Eligible	Registered
Class 50	13	13	36	36
Class 51	6	6	13	13

Service to Community

	Service Hours/Cadet	Dollar Value/ Hour	Total Value
Class 50	76	$22.95	$179,348
Class 51	77	$22.95	$185,666

Post-Residential Performance Status

	Graduated	Contacted	Placed	Education	Employment	Military	Multiple/ Other
Class 49							
Month 12	114	104	96	36	19	4	37
Class 50							
Month 1	103	101	87	41	17	1	29
Month 6	103	95	89	38	13	5	33
Month 12	103	100	95	32	22	4	40
Class 51							
Month 1	105	105	103	75	8	1	20
Month 6	105	101	97	43	14	1	34

Table A.36
Profile of Oregon Youth ChalleNGe Program

OREGON YOUTH CHALLENGE PROGRAM, ESTABLISHED 1999

Graduates Since Inception: 4,651	Program Type: Credit Recovery, High School Diploma, GED

Staffing

	Instructional	Cadre	Admin.	Case Managers	Recruiters	Other
Number employed	5	25	15	3	1	5

Funding

	Federal Funding	State Funding	Other Funding
Classes 50 and 51	$4,493,000	$1,507,609	$2,750

Residential Performance

	Dates	Applied	Entered Pre-ChalleNGe	Graduated	Received GED/HiSET	Received HS Credits	Received HS Diploma
Class 50	Jan 2018–Jun 2018	232	156	136	0	124	12
Class 51	Jul 2018–Dec 2018	232	157	139	0	129	10

Physical Fitness

	Push-Ups		1-Mile Run		BMI	
	Initial	Final	Initial	Final	Initial	Final
Class 50	27	51	09:07	07:24	26.3	25.8
Class 51	30	32	09:09	08:09	26.1	25.6

Responsible Citizenship

	Voting		Selective Service	
	Eligible	Registered	Eligible	Registered
Class 50	37	37	57	57
Class 51	22	22	72	72

Service to Community

	Service Hours/Cadet	Dollar Value/Hour	Total Value
Class 50	88	$25.40	$303,642
Class 51	94	$25.40	$333,289

Table A.36—Continued

Post-Residential Performance Status

	Graduated	Contacted	Placed	Education	Employment	Military	Multiple/ Other
Class 49							
Month 12	123	121	95	39	41	9	6
Class 50							
Month 1	136	136	124	82	27	1	14
Month 6	136	136	124	88	18	3	18
Month 12	136	136	107	59	20	6	26
Class 51							
Month 1	139	139	127	102	8	1	25
Month 6	139	139	106	70	12	4	16

Table A.37
Profile of Puerto Rico Youth ChalleNGe Academy

PUERTO RICO YOUTH CHALLENGE ACADEMY, ESTABLISHED 1999

Graduates Since Inception: 6,193 | **Program Type: High School Diploma, Credit Recovery**

Staffing

	Instructional	Cadre	Admin.	Case Managers	Recruiters	Other
Number employed	10	46	18	12	3	25

Funding

	Federal Funding	State Funding	Other Funding
Classes 50 and 51	$3,500,000	$1,166,666	$0

Residential Performance

	Dates	Applied	Entered Pre-ChalleNGe	Graduated	Received GED/ HiSET	Received HS Credits	Received HS Diploma
Class 50	May 2018–Oct 2018	295	261	217	0	0	215
Class 51	Oct 2018–Mar 2019	301	261	227	0	0	226

Table A.37—Continued

Physical Fitness

	Push-Ups		1-Mile Run		BMI	
	Initial	Final	Initial	Final	Initial	Final
Class 50	24	40	09:38	08:14	24.0	23.6
Class 51	21	40	10:11	08:10	24.5	23.3

Responsible Citizenship

	Voting		Selective Service	
	Eligible	Registered	Eligible	Registered
Class 50	51	51	42	42
Class 51	50	50	41	41

Service to Community

	Service Hours/Cadet	Dollar Value/Hr	Total Value
Class 50	40	$12.64	$109,715
Class 51	40	$12.64	$114,771

Post-Residential Performance Status

	Graduated	Contacted	Placed	Education	Employment	Military	Multiple/ Other
Class 49							
Month 12	216	216	180	110	42	1	27
Class 50							
Month 1	217	217	49	24	18	0	32
Month 6	217	217	173	117	34	5	29
Month 12	217	N/A	N/A	N/A	N/A	N/A	N/A
Class 51							
Month 1	227	227	43	10	26	0	24
Month 6	227	N/A	N/A	N/A	N/A	N/A	N/A

NOTES: N/A = not applicable; follow-up period has not occurred.

Table A.38
Profile of South Carolina Youth ChalleNGe Academy

SOUTH CAROLINA YOUTH CHALLENGE ACADEMY, ESTABLISHED 1998

Graduates Since Inception: 3,723 **Program Type: GED**

Staffing

	Instructional	Cadre	Admin.	Case Managers	Recruiters	Other
Number employed	8	27	15	5	2	8

Funding

	Federal Funding	State Funding	Other Funding
Classes 50 and 51	$2,744,000	$1,000,000	$0

Residential Performance

	Dates	Applied	Entered Pre-ChalleNGe	Graduated	Received GED/HiSET	Received HS Credits	Received HS Diploma
Class 50	Jan 2018–Jun 2018	96	93	60	24	0	0
Class 51	Jul 2018–Dec 2018	171	144	102	20	0	0

Physical Fitness

	Push-Ups		1-Mile Run		BMI	
	Initial	Final	Initial	Final	Initial	Final
Class 50	28	43	10:13	09:01	26.1	25.3
Class 51	29	*	10:00		25.2	*

Responsible Citizenship

	Voting		Selective Service	
	Eligible	Registered	Eligible	Registered
Class 50	15	15	12	8
Class 51	16	8	15	6

Service to Community

	Service Hours/Cadet	Dollar Value/Hr	Total Value
Class 50	46	$23.21	$64,060
Class 51	28	$23.21	$66,288

Table A.38—Continued

Post-Residential Performance Status

	Graduated	Contacted	Placed	Education	Employment	Military	Multiple/ Other
Class 49							
Month 12	103	93	17	3	9	2	3
Class 50							
Month 1	60	41	40	27	12	0	2
Month 6	60	18	18	10	8	0	0
Month 12	60	23	23	15	4	2	2
Class 51							
Month 1	102	50	50	36	10	2	2
Month 6	102	56	56	12	38	1	2

NOTE: * = did not report.

Table A.39
Profile of Volunteer Youth ChalleNGe Academy (Tennessee)

VOLUNTEER YOUTH CHALLENGE ACADEMY, ESTABLISHED 2017

Graduates Since Inception: 122	Program Type: High School Diploma, Credit Recovery, HiSET

Staffing

	Instructional	Cadre	Admin.	Case Managers	Recruiters	Other
Number employed	6	24	8	3	4	6

Funding

	Federal Funding	State Funding	Other Funding
Classes 50 and 51	$3,470,580	$1,156,860	$0

Residential Performance

	Dates	Applied	Entered Pre-ChalleNGe	Graduated	Received GED/ HiSET	Received HS Credits	Received HS Diploma
Class 50	Jan 2018–Jun 2018	104	65	46	*	*	*
Class 51	Jul 2018–Dec 2018	144	98	53	*	*	*

Table A.39—Continued

Physical Fitness

	Push-Ups		1-Mile Run		BMI	
	Initial	Final	Initial	Final	Initial	Final
Class 50	*	*	*	*	*	*
Class 51	*	*	*	*	*	*

Responsible Citizenship

	Voting		Selective Service	
	Eligible	Registered	Eligible	Registered
Class 50	10	0	9	0
Class 51	12	2	10	0

Service to Community

	Service Hours/Cadet	Dollar Value/ Hour	Total Value
Class 50	54	$22.67	$50,191
Class 51	41	$22.67	$50,191

Post-Residential Performance Status

	Graduated	Contacted	Placed	Education	Employment	Military	Multiple/ Other
Class 49							
Month 12	23	13	10	5	1	2	2
Class 50							
Month 1	46	39	36	7	13	0	16
Month 6	46	27	23	5	1	1	19
Month 12	46	5	0	0	0	0	3
Class 51							
Month 1	53	46	35	6	4	2	25
Month 6	53	29	11	2	1	1	2

NOTE: * = did not report.

Table A.40
Profile of Texas ChalleNGe Academy

TEXAS CHALLENGE ACADEMY, ESTABLISHED 2014

Graduates Since Inception: 448 **Program Type: Credit Recovery, High School Diploma, GED**

Staffing

	Instructional	Cadre	Admin.	Case Managers	Recruiters	Other
Number employed	5	21	5	5	4	2

Funding

	Federal Funding	State Funding	Other Funding
Classes 50 and 51	$2,550,000	$850,000	$150,000

Residential Performance

	Dates	Applied	Entered Pre-ChalleNGe	Graduated	Received GED/ HiSET	Received HS Credits	Received HS Diploma
Class 50	Jan 2018–Jun 2018	189	102	63	2	50	8
Class 51	Jul 2018–Dec 2018	216	155	87	4	66	11

Physical Fitness

	Push-Ups[a]		1-Mile Run		BMI	
	Initial	Final	Initial	Final	Initial	Final
Class 50	7	12	11:21	10:32	*	*
Class 51	35	38	10:46	10:50	25.8	27.6

Responsible Citizenship

	Voting		Selective Service	
	Eligible	Registered	Eligible	Registered
Class 50	10	10	7	7
Class 51	16	16	23	23

Service to Community

	Service Hours/Cadet	Dollar Value/ Hour	Total Value
Class 50	64	$25.10	$100,618
Class 51	51	$25.10	$109,117

Table A.40—Continued

Post-Residential Performance Status

	Graduated	Contacted	Placed	Education	Employment	Military	Multiple/ Other
Class 49							
Month 12	55	54	30	11	14	2	3
Class 50							
Month 1	63	49	26	18	5	0	3
Month 6	63	55	48	23	19	1	5
Month 12	63	50	43	20	13	7	3
Class 51							
Month 1	87	75	34	22	9	1	2
Month 6	87	80	70	18	37	7	8

NOTE: * = did not report.

[a]CLASS 50 reported pull-ups.

Table A.41
Profile of Virginia Commonwealth ChalleNGe Youth Academy

VIRGINIA COMMONWEALTH CHALLENGE YOUTH ACADEMY, ESTABLISHED 1994

Graduates Since Inception: 4,996 **Program Type: Credit Recovery, GED**

Staffing

	Instructional	Cadre	Admin.	Case Managers	Recruiters	Other
Number employed	8	35	13	4	3	14

Funding

	Federal Funding	State Funding	Other Funding
Classes 50 and 51	$3,824,572	$1,592,153	$211,787

Residential Performance

	Dates	Applied	Entered pre-ChalleNGe	Graduated	Received GED/ HiSET	Received HS Credits	Received HS Diploma
Class 50	Oct 2017–Feb 2018	233	167	104	45	42	0
Class 51	Mar 2018–Aug 2018	213	135	91	31	55	0

Table A.41—Continued

Physical Fitness

	Push-Ups		1-Mile Run		BMI	
	Initial	Final	Initial	Final	Initial	Final
Class 50	22	44	08:54	08:23	23.7	*
Class 51	14	40	10:17	08:35	24.7	*

Responsible Citizenship

	Voting		Selective Service	
	Eligible	Registered	Eligible	Registered
Class 50	25	25	46	46
Class 51	16	15	32	32

Service to Community

	Service Hours/Cadet	Dollar Value/Hour	Total Value
Class 50	105	$27.50	$300,300
Class 51	58	$27.50	$145,145

Post-Residential Performance Status

	Graduated	Contacted	Placed	Education	Employment	Military	Multiple/ Other
Class 49							
Month 12	116	36	31	6	13	2	10
Class 50							
Month 1	104	59	42	15	17	2	8
Month 6	104	31	23	7	9	1	7
Month 12	104	37	31	9	11	3	9
Class 51							
Month 1	91	91	91	62	11	3	15
Month 6	91	66	65	25	14	1	16

NOTE: * = did not report.

Table A.42
Profile of Washington Youth Academy

WASHINGTON YOUTH ACADEMY, ESTABLISHED 2009

Graduates Since Inception: 2,468 | **Program Type: Credit Recovery**

Staffing

	Instructional	Cadre	Admin.	Case Managers	Recruiters	Other
Number employed	6	30	13	6	2	13

Funding

	Federal Funding	State Funding	Other Funding
Classes 50 and 51	$3,800,000	$1,266,667	$2,140,065

Residential Performance

	Dates	Applied	Entered Pre-ChalleNGe	Graduated	Received GED/HiSET	Received HS Credits	Received HS Diploma
Class 50	Jan 2018–Jun 2018	264	166	137	0	137	0
Class 51	Jul 2018–Dec 2018	304	165	145	0	145	0

Physical Fitness

	Push-Ups		1-Mile Run		BMI	
	Initial	Final	Initial	Final	Initial	Final
Class 50	17	42	10:32	07:04	26.5	*
Class 51	20	38	11:02	07:14	26.4	*

Responsible Citizenship

	Voting		Selective Service	
	Eligible	Registered	Eligible	Registered
Class 50	37	37	61	61
Class 51	39	39	71	71

Service to Community

	Service Hours/Cadet	Dollar Value/Hr	Total Value
Class 50	56	$31.72	$241,897
Class 51	65	$31.72	$300,341

Table A.42—Continued

Post-Residential Performance Status

	Graduated	Contacted	Placed	Education	Employment	Military	Multiple/ Other
Class 49							
Month 12	141	141	110	63	33	13	1
Class 50							
Month 1	137	137	134	133	0	0	1
Month 6	137	136	131	124	5	0	2
Month 12	137	131	120	101	17	1	1
Class 51							
Month 1	145	145	141	139	0	0	2
Month 6	145	145	131	125	4	0	1

NOTE: * = did not report.

Table A.43
Profile of Wisconsin ChalleNGe Academy

WISCONSIN CHALLENGE ACADEMY, ESTABLISHED 1998

Graduates Since Inception: 3,811	Program Type: Credit Recovery, High School Diploma, GED

Staffing

	Instructional	Cadre	Admin.	Case Managers	Recruiters	Other
Number employed	4	21	8	4	4	2

Funding

	Federal Funding	State Funding	Other Funding
Classes 50 and 51	$3,495,603	$1,165,201	$0

Residential Performance

	Dates	Applied	Entered Pre-ChalleNGe	Graduated	Received GED/HiSET	Received HS Credits	Received HS Diploma
Class 50	Jan 2018–Jun 2018	246	140	100	41	0	59
Class 51	Jul 2018–Dec 2018	221	126	88	20	0	68

Table A.43—Continued

Physical Fitness

	Push-Ups		1-Mile Run		BMI	
	Initial	Final	Initial	Final	Initial	Final
Class 50	14	23	11:19	07:18	26.2	25.5
Class 51	20	41	09:18	07:56	25.4	25.4

Responsible Citizenship

	Voting		Selective Service	
	Eligible	Registered	Eligible	Registered
Class 50	23	23	55	55
Class 51	18	18	44	44

Service to Community

	Service Hours/Cadet	Dollar Value/ Hour	Total Value
Class 50	59	$25.12	$148,208
Class 51	78	$25.12	$172,424

Post-Residential Performance Status

	Graduated	Contacted	Placed	Education	Employment	Military	Multiple/ Other
Class 49							
Month 12	108	100	80	4	48	9	19
Class 50							
Month 1	100	100	60	2	46	0	25
Month 6	100	69	54	4	36	5	14
Month 12	100	20	15	0	12	1	3
Class 51							
Month 1	88	82	39	3	29	1	26
Month 6	88	14	12	0	8	1	1

Table A.44
Profile of Mountaineer ChalleNGe Academy (West Virginia)

MOUNTAINEER CHALLENGE ACADEMY, ESTABLISHED 1993	
Graduates Since Inception: 4,358	Program Type: High School Diploma

Staffing

	Instructional	Cadre	Admin.	Case Managers	Recruiters	Other
Number employed	8	35	19	6	3	2

Funding

	Federal Funding	State Funding	Other Funding
Classes 50 and 51	$4,500,000	$1,500,000	$0

Residential Performance

	Dates	Applied	Entered Pre-ChalleNGe	Graduated	Received GED/HiSET	Received HS Credits	Received HS Diploma
Class 50	Jan 2018–Jun 2018	358	174	139	0	21	118
Class 51	Jul 2018–Dec 2018	393	219	173	0	17	156

Physical Fitness

	Push-Ups		1-Mile Run		BMI	
	Initial	Final	Initial	Final	Initial	Final
Class 50	*	*	10:31	08:17	*	*
Class 51	24	41	10:10	07:51	*	25.1

Responsible Citizenship

	Voting		Selective Service	
	Eligible	Registered	Eligible	Registered
Class 50	35	35	31	31
Class 51	35	35	37	36

Service to Community

	Service Hours/Cadet	Dollar Value/ Hour	Total Value
Class 50	48	$22.29	$149,800
Class 51	49	$22.29	$190,314

Table A.44—Continued

Post-Residential Performance Status

	Graduated	Contacted	Placed	Education	Employment	Military	Multiple/ Other
Class 49							
Month 12	153	153	107	12	74	11	10
Class 50							
Month 1	139	138	30	1	26	3	2
Month 6	139	138	80	10	49	11	16
Month 12	139	138	88	8	59	14	8
Class 51							
Month 1	173	173	50	10	32	4	4
Month 6	173	173	99	10	65	18	3

* = did not report.

Table A.45
Profile of Wyoming Cowboy ChalleNGe Academy

WYOMING COWBOY CHALLENGE ACADEMY, ESTABLISHED 2005

Graduates Since Inception: 1,144 | **Program Type: Credit Recovery, HiSET**

Staffing

	Instructional	Cadre	Admin.	Case Managers	Recruiters	Other
Number employed	4	21	7	2	5	6

Funding

	Federal Funding	State Funding	Other Funding
Classes 50 and 51	$1,991,521	$1,576,442	$0

Residential Performance

	Dates	Applied	Entered Pre-ChalleNGe	Graduated	Received GED/HiSET	Received HS Credits	Received HS Diploma
Class 50	Jan 2018–Jun 2018	153	113	81	38	0	0
Class 51	Jul 2018–Dec 2018	58	50	34	22	0	0

Table A.45—Continued

Physical Fitness

	Push-Ups		1-Mile Run		BMI	
	Initial	Final	Initial	Final	Initial	Final
Class 50	24	37	08:21	07:41	*	*
Class 51	28	44	09:31	07:50	*	*

Responsible Citizenship

	Voting		Selective Service	
	Eligible	Registered	Eligible	Registered
Class 50	17	17	15	15
Class 51	3	3	3	3

Service to Community

	Service Hours/Cadet	Dollar Value/ Hour	Total Value
Class 50	47	$24.60	$100,434
Class 51	43	$24.60	$37,220

Post-Residential Performance Status

	Graduated	Contacted	Placed	Education	Employment	Military	Multiple/ Other
Class 49							
Month 12	60	60	13	7	4	1	1
Class 50							
Month 1	81	81	33	11	21	1	0
Month 6	81	32	32	12	16	4	0
Month 12	81	35	27	12	13	2	0
Class 51							
Month 1	34	21	5	2	2	1	0
Month 6	34	28	20	9	6	5	0

NOTE: * = did not report.

APPENDIX B

Additional Information About the TABE

Recent Changes to the TABE

The TABE 11/12 differs from previous versions of the test in several key ways:

- **Assessment Format and Length**. In prior versions of the TABE, ChalleNGe programs could choose between using the TABE Survey or the TABE Complete Battery test formats. The TABE Survey was a short-form assessment, based on 160 test items and with an administration time of two hours and eight minutes (2:08). The TABE Complete Battery was a long-form assessment, with 225 test items and with an administration time of three hours and 37 minutes (3:37) (DRC, 2016). By contrast, TABE 11/12 has only a single format, based on approximately 120 test items and with a maximum administration time of three hours and forty minutes (3:40) (DRC, 2019b).
- **Test Medium**. Prior to TABE 11/12, the assessment could be administered using paper and pencil, a Windows-based computer application, or through an online assessment platform. TABE 11/12 can be administered in either the paper-and-pencil or online-based formats. ChalleNGe programs will no longer be able to use computers without a live internet connection for TABE administration.
- **Content Areas**. The TABE 9/10 Survey format provided scores in the area of total reading, total mathematics, total language, and an overall composite score, Total Battery. The TABE 9/10 Complete Battery format assessed reading, mathematics computation, applied mathematics, and language, with optional tests for vocabulary, language mechanics, and spelling. The TABE 9/10 Complete Battery format also enabled the calculation of a composite Battery score. By contrast, TABE 11/12 covers three core content areas—reading, language, and mathematics—and does not provide an overall composite score.
- **Scoring**. This may be the area of most notable change for ChalleNGe programs. TABE 11/12 does not provide grade-equivalents for test takers.[1] Many ChalleNGe programs have relied on reporting the average change in cadet grade-equivalents from the start of the Residential Phase until the end of the Residential Phase. Consequently, programs will be required to adjust to new ways of examining cadet academic improvements during the course of the Residential Phase. TABE 11/12 will now provide only a scale score, ranging from 300 to 800. DRC has released scale score guidance, aligning the new scale scores to

[1] A grade-equivalent equated a test taker's performance to that of a student in the 50th percentile of achievement in a particular grade and month. For example, a grade equivalent of 8.7 suggests that a test taker's performance is equivalent to a student performing at the 50th percentile of achievement for someone who is in the seventh month of eighth grade.

grade levels (DRC, undated) to ease the transition to focusing on the scale scores in the TABE 11/12.

Table B.1
TABE Reading Grade Levels and Associated Scale Scores, by TABE 9/10 and 11/12

Grade Level	TABE 9/10 Scale Scores	TABE 11/12
0	160–294	300–371
1	300–367	372–441
2	368–428	442–471
3	429–460	472–500
4	461–486	501–518
5	487–517	519–535
6	518–536	536–549
7	537–549	550–562
8	553–565	563–575
9	567–581	576–596
10	582–595	597–616
11	598–606	617–709
12	608–612	710–800
12.9+	614–812	—

NOTES: On TABE 9/10, cadets could reach the ceiling (i.e., top possible score) on an assessment if they were tested using the wrong form. In such cases, cadets would receive a grade-equivalent marked with a "+" to indicate that the cadet hit the ceiling and her true performance level was not identified. We have removed, with the exception of the 12.9+ grade equivalent, these grade-equivalent/scale score ranges from this table to ease understanding. TABE 11/12 will also indicate where a cadet was assessed using the wrong form of the TABE assessment (i.e., an assessment that was too hard or too easy). Administering the TABE locator before both the pre- and post-TABE assessments mitigates this issue by identifying the appropriate test form a cadet should be given for each test subject.

References

Bloom, Dan, Alissa Gardenhire-Crooks, and Conrad Mandsager, *Reengaging High School Dropouts: Early Results of the National Guard Youth ChalleNGe Program Evaluation*, New York: MDRC, February 2009.

Bozick, Robert, and Keith MacAllum, "Does Participation in a School-to-Career Program Limit Educational and Career Opportunities?" *Journal of Career and Technical Education*, Vol. 18, No. 2, 2002.

Bozick, Robert, Gabriella C. Gonzalez, Serafina Lanna, and Monica Mean, *Preparing New York City High School Students for the Workforce: Evaluation of the Scholars at Work Program*, Santa Monica, Calif.: RAND Corporation, RR-2488-NYCCEO, 2019. As of November 11, 2019:
https://www.rand.org/pubs/research_reports/RR2488.html

Burns, B., S. Phillips, H. Wagner, R. Barth, D. Kolko, Y. Campbell, and J. Yandsverk, "Mental Health Need and Access to Mental Health Services by Youths Involved with Child Welfare: A National Survey," *Journal of the American Academy of Child and Adolescent Psychiatry*, Vol. 43, No. 8, 2004, pp. 960–970.

Castellano, Maria, Kristen E. Sundell, Laura T. Overman, George B. Richardson, and James R. Stone, *Rigorous Tests of Student Outcomes in CTE Programs of Study: Final Report*, Louisville, Ky.: National Research Center for Career and Technical Education, April 2014.

Comprehensive Adult Student Assessment System, "Study of the CASAS Relationship to GED 2002," San Diego, Calif., research brief, June 2003. As of November 8, 2017:
https://www.casas.org/docs/default-source/research/download-what-is-the-relationship-between-casas-assessment-and-ged-2002-.pdf?sfvrsn=5?Status=Master

———, "Study of the CASAS Relationship to GED 2014," San Diego, Calif., research brief, March 2016. As of November 8, 2017:
https://www.casas.org/docs/default-source/research/study-of-the-casas-relationship-to-ged-2014.pdf

Constant, Louay, Jennie W. Wenger, Linda Cottrell, Wing Yi Chan, and Kathryn A. Edwards, *National Guard Youth ChalleNGe: Program Progress in 2017–2018*, Santa Monica, Calif.: RAND Corporation, RR-2907-OSD, 2019. As of November 8, 2019:
https://www.rand.org/pubs/research_reports/RR2907.html

Data Recognition Corporation, "TABE 11 & 12 Grade Range Scale Score Guidance," undated. As of November 14, 2019:
https://tabetest.com/PDFs/TABE_GradeRange.pdf

———, "Discover TABE 9&10," promotional brochure, 2016. As of November 19, 2019:
https://tabetest.com/PDFs/TABE_9_10_Brochure.pdf

———, "New TABE 11 & 12 Receives NRS Approval, Leads Market as First and Only Updated Assessment of Adult Education," press release, October 17, 2017. As of November 11, 2019:
https://tabetest.com/wp-content/uploads/2017/10/TABE_11_12_Press_Release_10_17_17_FINAL_Version.pdf

———, *TABE Tests of Adult Basic Education: Adult Education Solutions*, Maple Grove, Minn., 2019a. As of November 11, 2019:
https://tabetest.com/PDFs/TABE_Overview_Brochure.pdf

———, "TABE 11&12: Maximum Allowable Testing Times," fact sheet, July 19, 2019b. As of November 19, 2019:
http://www.tabetest.com/PDFs/TABE_11_12_Max_Testing_Times.pdf

DRC—*See* Data Recognition Corporation.

DuBois, D. L., N. Portillo, J. E. Rhodes, N. Silverthorn, and J. C. Valentine, "How Effective Are Mentoring Programs for Youth? A Systematic Assessment of the Evidence," *Psychological Science in the Public Interest*, Vol. 12, No. 2, 2011, pp. 57–91.

Dweck, C. S., *Mindset: The New Psychology of Success*, New York: Random House, 2006.

Garringer, M., J. Kupersmidt, J. Rhodes, R. Stelter, and T. Tai, *Elements of Effective Practice for Mentoring: Research-Informed and Practitioner-Approved Best Practices for Creating and Sustaining Impactful Mentoring Relationships and Strong Program Services*, 4th ed., Boston, Mass.: MENTOR: The National Mentoring Partnership, 2015. As of November 11, 2019:
https://www.mentoring.org/program-resources/elements-of-effective-practice-for-mentoring/

Hamilton Fish Institute on School and Community Violence and The National Mentoring Center at Northwest Regional Educational Laboratory, *Training New Mentors: Effective Strategies for Providing Quality Youth Mentoring in Schools and Communities*, Washington, D.C., 2007.

Hemelt, Steven W., Matthew A. Lenard, and Colleen G. Paeplow, "Building Bridges to Life After High School: Contemporary Career Academies and Student Outcomes," *Economics of Education Review*, Vol. 68, 2019, pp. 161–178.

Independent Sector, "Value of Volunteer Time," webpage, undated. As of December 7, 2017:
http://www.independentsector.org/resource/the-value-of-volunteer-time/

Karp, Melinda Mechur, Juan Carlos Calcagno, Katherine Hughes, Dong Wook Jeong, and Thomas Bailey, *The Postsecondary Achievement of Participants in Dual Enrollment: An Analysis of Student Outcomes in Two Sites*, St. Paul, Minn.: National Research Center for Career and Technical Education, University of Minnesota, 2007.

Kemple, James, *Career Academies: Long-Term Impacts on Work, Education, and Transitions to Adulthood*, New York: MDRC, 2008.

Knowlton, Lisa Wyatt, and Cynthia C. Phillips, *The Logic Model Guidebook: Better Strategies for Great Results*, Thousand Oaks, Calif.: Sage Publications, 2009.

Lindholm-Leary, Kathryn, and Gary Hargett, *Evaluator's Toolkit for Dual Language Programs*, Sacramento, Calif.: California Department of Education, December 2006.

MENTOR, The National Mentoring Partnership, "About MENTOR," webpage, undated. As of November 12, 2019:
https://www.mentoring.org/our-work/about-mentor/

Millenky, Megan, Dan Bloom, and Colleen Dillon, *Making the Transition: Interim Results of the National Guard Youth ChalleNGe Evaluation*, New York: MDRC, May 2010.

Millenky, Megan, Dan Bloom, Sara Muller-Ravett, and Joseph Broadus, *Staying on Course: Three-Year Results of the National Guard Youth ChalleNGe Evaluation*, New York: MDRC, 2011.

National Guard Youth ChalleNGe, homepage, undated. As of November 11, 2019:
https://youth.gov/federal-links/national-guard-youth-challenge-program

———, *2015 Performance and Accountability Highlights*, Arlington, Va.: National Guard Bureau, December 2015.

National Reporting Service for Adult Education, "NRS Test Benchmarks for Educational Functioning Levels," Washington, D.C., 2015.

National Research Council, *The Importance of Common Metrics for Advancing Social Science Theory and Research: A Workshop Summary*, Washington, D.C.: National Academies Press, 2011.

Olsen, M., *Guide to Administering TABE (Tests of Adult Basic Education): A Handbook for Teachers and Test Administrators*, Little Rock, Ark.: Arkansas Department of Career Education, 2009. As of January 13, 2017:
http://ace.arkansas.gov/adulteducation/documents/tabehandbook2009.pdf

Perez-Arce, Francisco, Louay Constant, David S. Loughran, and Lynn A. Karoly, *A Cost-Benefit Analysis of the National Guard Youth ChalleNGe Program,* Santa Monica, Calif.: RAND Corporation, TR-1193-NGYF, 2012. As of October 17, 2017:
http://www.rand.org/pubs/technical_reports/TR1193.html

Pimental, S., *College and Career Readiness Standards for Adult Education*, Washington, D.C.: U.S. Department of Education Office of Vocational and Adult Education, 2017. As of November 11, 2019:
http://lincs.ed.gov/publications/pdf/CCRStandardsAdultEd.pdf

Porche, M. V., L. R. Fortuna, J. Lin, and M. Alegria, "Childhood Trauma and Psychiatric Disorders As Correlates of School Dropout in a National Sample of Young Adults," *Child Development*, Vol. 82, No. 3, 2011, pp. 982–998.

Price, Hugh, "Foundations, Innovation and Social Change: A Quixotic Journey Turned Case Study," working paper presented during practitioner residency, Rockefeller Foundation Bellagio Center, 2010. As of October 17, 2017:
http://cspcs.sanford.duke.edu/sites/default/files/Foundations%20Innovation%20and%20Social%20Change.pdf

Project for Education Research That Scales and MENTOR, "Growth Mindset for Mentors," online toolkit, undated. As of November 15, 2019:
https://www.mindsetkit.org/growth-mindset-mentors

Shakman, Karen, and Sheila M. Rodriguez, *Logic Models for Program Design, Implementation, and Evaluation: Workshop Toolkit*, Washington, D.C.: U.S. Department of Education, 2015. As of October 17, 2017:
http://files.eric.ed.gov/fulltext/ED556231.pdf

Skowyra, K. R. and J. J. Cocozza, *Blueprint for Change: A Comprehensive Model for the Identification and Treatment of Youth with Mental Health Needs in Contact with the Juvenile Justice System*, Delmar, N.Y.: National Center for Mental Health and Juvenile Justice and Policy Research Associates, Inc., 2006.

TABE— *See* Tests of Adult Basic Education.

"Tests Determined to Be Suitable for Use in the National Reporting System for Adult Education," U.S. Department of Education Notice, 83 FR 47910, 2018-20590, September 21, 2018. As of November 11, 2019:
https://www.federalregister.gov/documents/2018/09/21/2018-20590/
tests-determined-to-be-suitable-for-use-in-the-national-reporting-system-for-adult-education

Tests of Adult Basic Education, "States Using TABE," webpage, undated. As of December 7, 2017:
http://tabetest.com/resources-2/states-using-tabe/

U.S. Code, Title 32, Chapter Five, Section 509, National Guard Youth ChalleNGe Program of Opportunities for Civilian Youth, Part K, "Report."

U.S. Department of Defense Instruction 1025.8, National Guard ChalleNGe Program, Washington, D.C., March 20, 2002.

U.S. Department of Education, Institute of Education Sciences, National Center for Education Statistics, "High School Longitudinal Study of 2009 (HSLS:09), Base-year, 2013 Update, and High School Transcript File," 2019a. As of November 21, 2019:
https://nces.ed.gov/surveys/ctes/tables/h185.asp

———, "High School Longitudinal Study of 2009 (HSLS:09), Base-year, 2013 Update, and High School Transcript File," 2019b. As of November 21, 2019:
https://nces.ed.gov/surveys/ctes/tables/h186.asp

U.S. Department of Education, Office of Career, Technical, and Adult Education, Division of Adult Education and Literacy, *Implementation Guidelines: Measures and Methods for the National Reporting System for Adult Education*, Washington, D.C., 2016.

U.S. Department of Labor, U.S. Department of Commerce, U.S. Department of Education, and U.S. Department of Health and Human Services, "What Works in Job Training: A Synthesis of the Evidence," 2014.

Visher, M. G., and Stern, D., *New Pathways to Careers and College: Examples, Evidence, and Prospects*, New York: MDRC, 2015.

Wagner, M., "Youth with Disabilities Leaving Secondary School," in *Changes over Time in the Early Post School Outcomes of Youth with Disabilities: A Report of Findings from the National Longitudinal Transition Study (NLTS) and the National Longitudinal Transition Study-2 (NLTS2)*, Menlo Park, Calif.: SRI International, 2005, pp. 2.1–2.6.

Wenger, Jennie W., Louay Constant, and Linda Cottrell, *National Guard Youth ChalleNGe: Program Progress in 2016–2017*, Santa Monica, Calif.: RAND Corporation, RR-2276-OSD, 2018. As of November 8, 2019: https://www.rand.org/pubs/research_reports/RR2276.html

Wenger, Jennie W., Louay Constant, Linda Cottrell, Thomas E. Trail, Michael J. D. Vermeer, and Stephani L. Wrabel, *National Guard Youth ChalleNGe: Program Progress in 2015–2016*, Santa Monica, Calif.: RAND Corporation, RR-1848-OSD, 2017. As of November 7, 2019: https://www.rand.org/pubs/research_reports/RR1848.html

Wenger, Jennie W., and Apriel K. Hodari, *Predictors of Attrition: Attitudes, Behaviors, and Educational Characteristics*, Washington, D.C.: Center for Naval Analyses, CRM D0010146.A2/Final, 2004.

Wenger, Jennie, Cathleen McHugh, and Lynda Houck, *Attrition Rates and Performance of ChalleNGe Participants over Time*, Washington, D.C.: Center for Naval Analyses, CRM D0013758.A2/Final, 2006.

Wenger, Jennie, Cathleen McHugh, Seema Sayala, and Robert Shuford, *Variations in Participants and Policies Across ChalleNGe Programs*, Arlington, Va.: Center for Naval Analyses, CRM D0016643.A2/Final, 2008.

West Virginia Department of Education, *Correlation Between Various Placement Instruments for Reading, Language/Writing, Mathematics, Elementary Algebra*, Charleston, W.Va., undated. As of November 8, 2017: https://wvde.state.wv.us/abe/documents/CorrelationBetweenVariousPlacementInstruments.pdf